THE VOLCANIC ROCKS OF THE LAKE DISTRICT:
A GEOLOGICAL GUIDE TO THE CENTRAL FELLS

Other Macmillan titles of related interest

D. V. Ager, *The Nature of the Stratigraphical Record*
J. R. Haynes, *Foraminifera*
C. S. Hutchison, *Economic Deposits and their Tectonic Setting*
H. H. Read and Janet Wilson, *Introduction to Geology:*
                                      Volume 1 Principles
                                      Volume 2 Earth History (Early Stages of Earth History)
                                      Volume 3 Earth History (Late Stages of Earth History)

The Volcanic Rocks of the Lake District

A Geological Guide to the Central Fells

F. MOSELEY
Reader in Geology,
Department of Geological Sciences,
University of Birmingham

First published 1983 by
THE MACMILLAN PRESS LTD
London and Basingstoke
Companies and representatives
throughout the world

Printed in Great Britain by
Unwin Brothers Limited
The Gresham Press, Old Woking, Surrey

Typeset by RDL Artset, Sutton, Surrey

ISBN 0 333 34977 6

Contents

Preface

With few exceptions the Cumbrian mountains are formed from volcanic rocks that are remnants of a volcano active during the Ordovician Period, 450 million years ago. These rocks account for the steep crags of the central fells between Keswick, Coniston, Ambleside and Haweswater. This is the main tourist area of the Lake District to which many thousands of visitors are attracted each year by the magnificent lake and mountain scenery, a scenery that depends to a great extent on the geology. It is understandable therefore that walkers, climbers and even motorists should be curious about the geological origins of the mountains and the valleys, and it is my firm belief that some knowledge of the geology greatly enhances appreciation of the scenery.

This book is an attempt to explain the geology of this beautiful area so that the rocks can be identified and the nature of the ancient volcano understood. It is intended to appeal to those for whom geology is a hobby: to the walker whose curiosity is aroused by the rocky crags and the valleys, to students who have already been introduced to the subject at school and university, and to their teachers who will be experienced geologists, perhaps with considerable knowledge of the Lake District. In order to cater for such a diverse audience I have attempted to explain the geology in simple terms in the early chapters, and then to go into more detail in the excursion itineraries. There are numerous illustrations that should enable the reader to relate distant views of mountainsides to the geology, and to identify the varieties of rocks encountered along the walking routes across the fells.

I wish to acknowledge continuous encouragement from my wife, and the assistance I have received from the Geological Department drawing office at Birmingham University.

Finally I would apologise for my indiscriminate use of miles, kilometres, metres and feet. My excuse is that we should all be bilingual!

1

Introduction

The steepest and most spectacular crags and mountains of the central Lake District are nearly all of volcanic origin and represent the much eroded remnant of an ancient volcano active 450 million years ago during the Ordovician Period (figures 1 and 2). At this time Earth would have been unrecognisable to us; indeed it must have been like a distant planet from a far galaxy with the land masses completely devoid of life — it was still to be 100 million years before land plants became established. Moreover the shapes and positions of oceans and continents were entirely different from those of today. These conclusions are made possible by the great variety and detail of research in the geological sciences, which is increasing enormously year by year. This work has included the laboratory examination of rocks, numerous specimens of which have been inspected under the microscope and chemically analysed, so permitting comparisons to be made with similar rocks being formed at the present time. Studies of marine fossils from Ordovician rocks of the same age as the volcanic rocks of the Lake District have also yielded a great deal of information about the geographical conditions existing during the Ordovician Period. Determination of the magnetic properties of the rocks have enabled the latitudes at which they originated (palaeomagnetism) to be determined, and their ages have been calculated using methods involving radioactive decay. Other geophysical investigations including calculations of minor variations in the value of gravity (Bott, 1978) and

studies of seismic waves generated by explosives have enabled the nature and structure of rocks as much as 30 km below the land surface to be determined. These studies have been brought together in the modern ideas of plate tectonics outlined below which have made it possible to reconstruct the positions of the former oceans and land masses and to understand the geography of Ordovician times.

Figures 3 and 4 show the geography during the Ordovician when the 'Lake District' was part of the north-western margin of an ancient 'Europe', at a latitude of 25° south, and Scotland was on the opposite side of a wide ocean (Iapetus) forming part of the ancient 'America' (Moseley, 1978, 1982). The Iapetus Ocean was not in any way related to the present Atlantic, and in fact its structural relationships to the surrounding continents were quite different to those of the Atlantic Ocean. The modern Atlantic has 'passive margins'; that is the flanking continents of Europe and Africa and the adjacent ocean are part of the same 'plate' with no differential movement between continent and ocean crust (figure 5), and hence there is very little earthquake or volcanic activity. The ocean margins of the Iapetus were quite different, rather like the modern Pacific margins where oceanic crust descends beneath continental crust along an inclined plane (a subduction or Benioff zone) as shown on figure 5. This is referred to as an active or destructive plate margin. The differential movement generates friction and heat, and violent earthquakes and volcanicity are

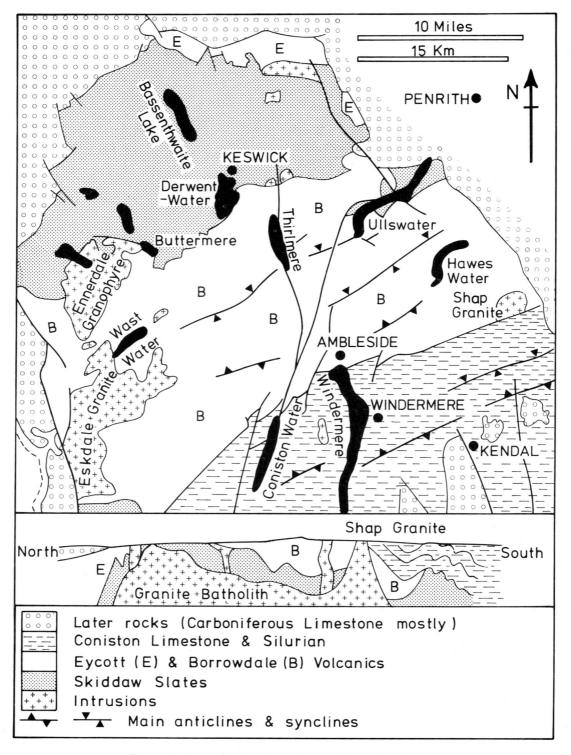

Figure 1   Generalised geological map of the Lake District

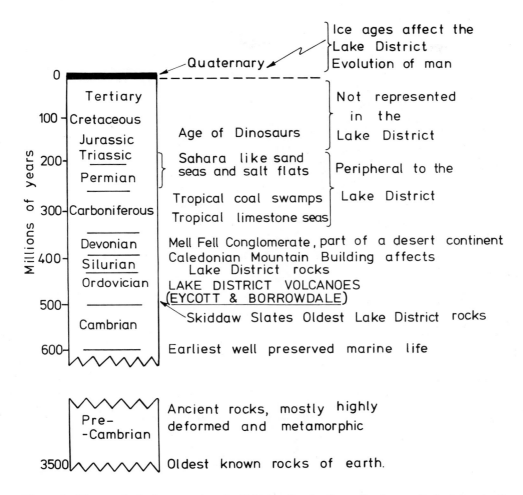

Figure 2 The geological succession in Britain. Rocks from the lower Ordovician to the Triassic are well represented in the Lake District and surrounding regions

commonplace. Oceans are not permanent features of Earth's surface: the modern Atlantic opened up in comparatively recent (Cretaceous) times, 100 million years ago, and still becomes wider each year, whereas the Iapetus Ocean was initiated rather more than 600 million years ago in the late Precambrian and was finally destroyed 400 million years ago at the end of the Silurian Period when the two ancient continents came together in 'collision' (figure 6). This was the last phase of the Caledonian Orogeny — mountain-building movements during which the rocks of the Lake District were severely folded and fractured (faulted). It

resulted also in the formation of a new continent (Eur-America) which extended from the Rockies to the Urals.

This book deals with the time from 500 to 400 million years ago and concerns the Ordovician volcano and its subsequent deformation during the Caledonian Orogeny. These events were long ago, and the many changes since then are outlined below so that Ordovician and Caledonian events can be seen in perspective.

Much of the rock variety seen in Britain is a result of steady northerly drift from one climatic zone to another. In Ordovician times the 'Lake

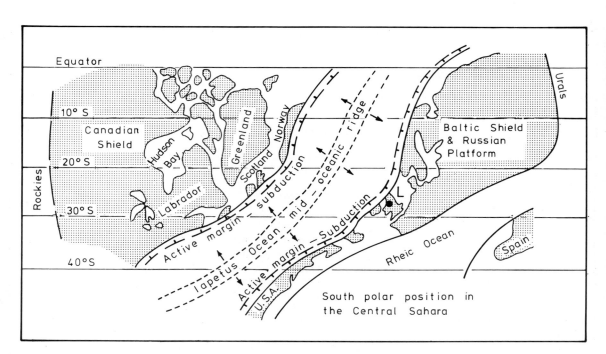

Figure 3 **Geography of the Iapetus Ocean region during the lower Ordovician. North America, including Scotland, the north of Ireland and most of Norway is shown separated from England, Wales, the south of Ireland and the rest of north-west Europe by the Iapetus Ocean. This ocean contracted and closed towards the end of the Ordovician period and no longer exists**

District' was in a southern hemisphere subtropical environment. It became part of a southern hemisphere mountain desert during the Devonian Period, 400 to 350 million years ago, with boulder beds deposited in intermontane basins (the Mell Fell Conglomerate), and must have been rather like present-day South Arabia. The region then drifted northwards through the tropical seas of the Carboniferous Limestone and the tropical swamps of the Coal Measures (250 million years ago), to northern hemisphere hot Sahara-like deserts in the Permo-Trias when the gypsum of the Vale of Eden and the thick salt deposits of other parts of Britain accumulated (200 million

years ago). During the Cretaceous Period about 100 million years ago, the ancient continent of Eur-America was breaking up and the modern Atlantic Ocean was opening. Finally, continued northerly drift brought the region into cool temperate latitudes, and some 2 million years ago Britain came within the sphere of the Quaternary ice age, with glaciers sweeping as far south as Bristol. This ice age is by no means over; we are merely indulging in an interglacial warmer phase. It is also a period of interest to *Homo Sapiens* since, although it represents only a moment of geological time, it has seen the evolution of our species.

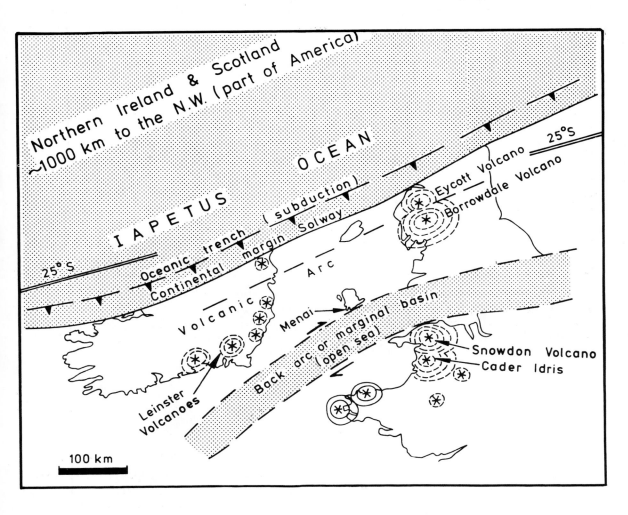

Figure 4 A reconstructed environment for the Ordovician volcanoes on the southern side of the Iapetus Ocean. It is suggested that northern England was separated from Wales by open sea, and that these two regions came together during later earth movements

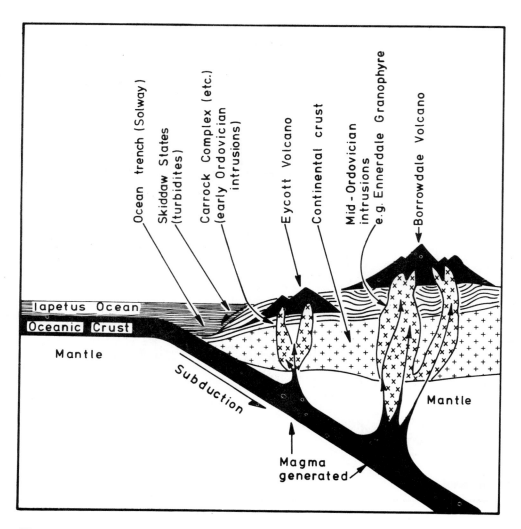

Figure 5 A section from the Iapetus Ocean across the lower Ordovician continental margin. The continent is shown over-riding oceanic crust that is subducted beneath it. The resulting high temperatures generated magmas that rose through the continental crust to form large intrusions (the Eskdale Granite etc.), eventually reaching the surface as the Eycott and Borrowdale Volcanoes. It seems likely that the magma chambers of the intrusions would have remained active after volcanicity ceased. This section is similar to one that has been constructed from the Pacific to the Cascade Range of north-west America (see figure 8)

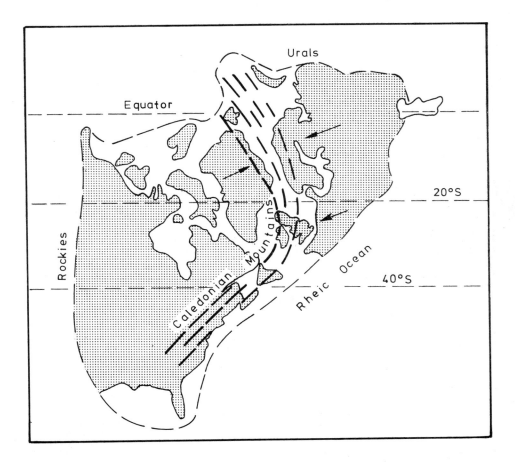

Figure 6 A new continent was formed during the final ocean closure and continental collision, which occurred at the end of the Silurian Period. Although the Iapetus Ocean had virtually disappeared millions of years earlier (towards the end of the Ordovician), the final impact of collision was delayed until this time. A new continent of Eur-america then came into being, and the high Caledonian mountain chain was raised above what became the Old Red Sandstone desert

2

Comparison between Modern Volcanoes and the Ordovician Volcano of the Lake District

An excellent way to understand the Ordovician volcanicity of the Lake District, Snowdonia and elsewhere in Britain is to compare it with similar volcanic activity of the present day. Each eruption is now documented as it takes place and the geographical and geological settings are known in relation to major structures of Earth such as active continental margins and other plate boundaries (see the introductory chapter, figure 5 and below). The environment of the older volcanics will not be so clear of course; they have been deformed and altered by later geological events and perhaps partly concealed by younger rocks, making them more difficult to interpret. It must be understood that the volcanoes of our present-day world occur in many different geological environments, most of which have little in common with the Ordovician environment of the Lake District. For example a great volume of basaltic lava rises from the mantle (see figures 5 and 7) along mid-ocean ridges; much of it erupted beneath the sea but some rose above the surface in areas such as Iceland. Other basaltic volcanoes occur as isolated oceanic centres away from these ridges (Hawaii), and there are related but more alkali-rich varieties associated with continental rifts such as in East Africa and the Red Sea. These latter are the mid-oceanic ridges of the future. The environments that most closely resemble that of the Lake District however are those with active or destructive continental margins and island arcs where oceanic crust descends beneath continental crust along subduction zones (figure

5). Volcanic chains of this type are to be found in Japan, the West Indies, New Zealand, the Andes and the Cascades of north-west America, and have been well documented. The movement between oceanic and continental crust results in earthquakes and generates heat which melts the rock at depth. The magma rises through the continental crust and is contaminated by it, eventually forming volcanoes with a mean composition of andesite (figure 7). The andesitic and related rocks of this environment are quite different mineralogically and chemically to the rocks of other volcanic environments, most of which are of basaltic composition, but details of these differences are outside the scope of this book.

To illustrate the nature of the Ordovician volcanoes of the Lake District I have compared them with the north-west American Cascades with which I am also familiar (figure 8). In this region a string of active volcanoes extends for 1000 miles from northern California to British Columbia, each major centre being about 30 miles from the next with the highest cone, Mount Rainier, over 14 000 feet above sea level. Eruptions are not common within the lifetime of one human being, indeed there have been only two this century (Lassen Peak from 1914 to 1917 and Mount St. Helens in 1980 and 1981 (figure 9)), but given a span of say 100 000 years (only a short time compared with the 50 million years of Ordovician volcanism in Britain) there has been a great deal of activity. Some of the recent eruptions have been of astounding violence; for example one of the largest was the

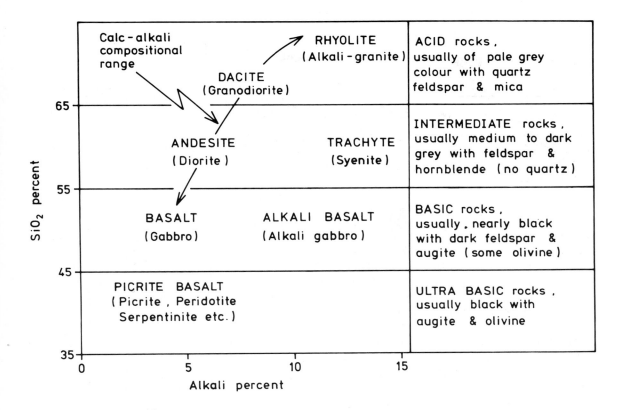

Figure 7 **A classification of igneous rocks with emphasis on the volcanic rocks found in the Lake District and similar environments**

disintegration of Mount Manzama and the formation of Crater Lake 8000 years ago, whilst the 1980 Mount St. Helens eruption although not on such a large scale was certainly impressive and caught the imagination of the world. There is no doubt that the Lake District volcanoes (really two volcanic complexes operating at different times) would have resembled Cascade volcanoes in the nature of the eruptions, the volcanic products and the time intervals between eruptions, the only difference being that they were more commonly surrounded by water and a greater proportion of the ash was therefore deposited in water.

Like the Cascades there was a string of volcanoes extending across Britain during the Ordovician (Figure 4), and like the Cascades new eruptions would have destroyed parts of the existing volcanoes. During the hundreds of years between one eruption and the next, normal river erosion would have removed vast quantities of the loose ash from the slopes of the volcano, especially during Ordovician times when the land had not yet been colonised by plants. Indeed the first land plants did not appear until the Devonian Period (figure 2).

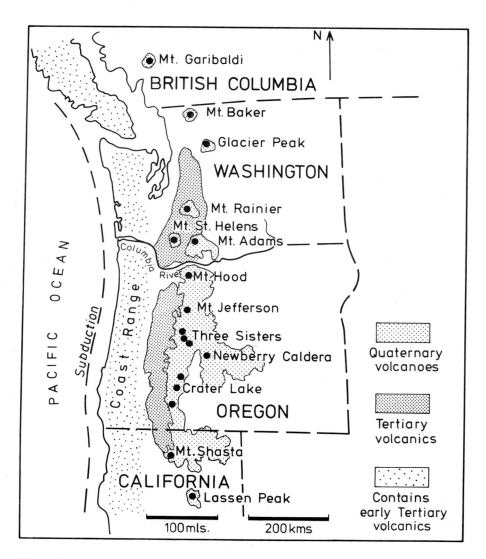

Figure 8 The Cascade Range of north-west America extending from Mount Garibaldi to Lassen Peak can serve as a present-day model for the Ordovician volcanoes of Britain

Figure 9 Mount St. Helens in the Cascades from the west, before and after the eruption of 1980. More than 1000 feet of rock was blown from the summit in one gigantic explosion on 18 May 1980. This volcano is 10 miles in diameter, whilst several others, for example Mount Rainier farther north, are much larger. These dimensions are comparable to those of the former Borrowdale volcano and the compositions of the lavas and tuffs are also similar

3

Field Identification of Lake District Volcanic Rocks

The magma of the volcanoes of island arcs and continental margins originates along subduction zones. It is initially derived from the mantle and oceanic crust and is low in silica, but as it rises through the granitic rocks of the continental crust it becomes contaminated by these silica-rich rocks and assumes an average composition of andesite (figure 7). There is however considerable chemical variation from basalt at one extreme to rhyolite at the other depending on local conditions, and there is also considerable variations in the form of the extrusive material, from lava flows to explosive fragmental material and volcanic mudflows. The silica-poor basalts and basaltic andesites form the more fluid magmas and are frequently found as lava flows, although breccia (large fragments) and tuff (finer-grained ash), formed when rocks disintegrated by explosion, are also common. The opposite is true of the more viscous silica-rich rhyolite and dacite magma which frequently occur in the form of disastrously explosive ignimbrite (eruptions of incandescent ash) although lavas of this composition are also common. All these rocks are characteristic of the continental margin and island arc volcanoes which have been mentioned above. In the north-western American Cascades, which I have compared with the Lake District, there are basalt and basaltic andesite lavas just over 2000 years old (Sisters). Crater Lake, which is 5 miles in diameter and 4000 feet deep from crest to floor, was formed 8000 years ago by a gigantic explosion that disintegrated 4 cubic miles of rock, the explosive magma being of rhyo-dacite composition (it is asserted by Indian legend that this was the result of two fighting gods who were rather annoyed with each other!). Most recently the highly explosive rhyo-dacitic eruption of Mount St. Helens in 1980 generated so much heat that a hot ash-flow blast brought lakes and rivers to boiling point. About 1 cubic kilometre of rock was pulverised to ash which, mixed with volcanic gas and steam, mushroomed into the stratosphere; glaciers melted and valleys were filled with volcanic mudflows such as that which devastated the Toutle River west of Mount St. Helens. Although this was merely one eruption, it gave many different volcanic forms: an initial ground-hugging high-temperature ignimbrite blast which destroyed everything in its path, an enormous volume of fine ash fanning east within the jet stream to be deposited over an area of thousands of square miles, the mudflows of the Toutle River, and less than a year later a large andesitic lava dome forming in the crater. All these deposits were virtually contemporaneous and certainly would be so in relation to geological time. Other similar events are likely to occur at intervals of from a hundred to several thousand years, gradually constructing and destroying such a volcano. It is easy therefore to appreciate the difficulties in interpreting the history of an ancient volcano such as that of the Lake District (Millward *et al.*, 1978). However a principal object of this book is to enable the non-specialist to identify some of the rocks as well as to appreciate some of the events of the times. Outline descriptions of Lake District rocks are therefore necessary and are given under the principal headings of lavas and pyroclastic rocks. It is preferable to examine these rocks on the exposed faces of the crags where weathering often picks out the texture to perfection. Freshly broken surfaces yield other information if examined with a hand lens, but indiscriminate chipping of the rock must be avoided; this leaves ugly marks which are visible for years. Specimens should be taken from outcrops that are not visible to others and very good material can be collected from scree.

LAVAS

Basalt and basaltic andesite

It is not possible to distinguish between true basalt and basaltic andesite without chemical analysis but the group as a whole can be identified in the field. These rocks are usually dark grey to blue grey and very fine grained so that individual crystals of the ground mass cannot be seen. However there are often larger crystals (phenocrysts) which are easily seen; these are usually black (pyroxene) or pale grey (feldspar) (figure 10). Many outcrops are

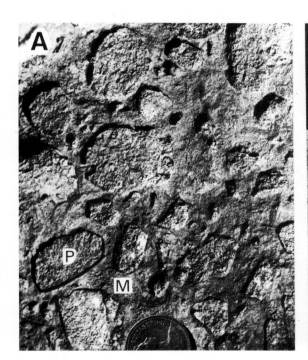
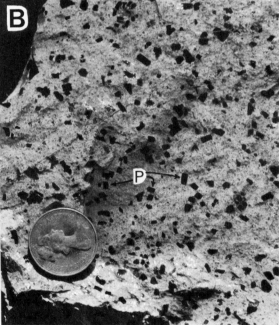

Figure 10 Porphyritic lava with large crystals (phenocrysts) set in a fine-grained crystalline matrix. (A) Basaltic andesite from Eycott Hill (NY 385 295) with large feldspar phenocrysts (P) set in a fine matrix (M). (B) Porphyritic andesite, Bramcrag, Vale of St. John (NY 320 220, excursion C), with rectangular black phenocrysts (P) in a fine-grained matrix. The phenocrysts were formerly pyroxene (augite), but have been altered to chlorite; but remember that such detailed identifications of minerals cannot be made in the field and depend on a microscopic study and geochemical analysis

cut by close spaced fractures (flow joints) which may be less than 1 cm apart, and since they formed parallel to the flowing lava surfaces they serve to give the angle of dip of the strata. It should be appreciated that these rocks were folded by the Caledonian mountain-building movements more than 100 million years after their formation (see below) and their present angle of inclination does not represent that at which they were originally formed. Apart from the flow joints other volcanic structures include flow brecciation and amygdaloidal texture, but these are both better displayed by andesite lava and will be described under that heading.

Andesite lava

This is the most common of the Lake District lavas and is richer in silica and lower in iron and magnesium than the basalts. It is generally medium grey and very fine grained but with a scatter of phenocrysts (see above) which are usually pale grey feldspar. Flow joints similar to those in basalts are common, sometimes contorted into flow-folded structures as on Haystacks (figure 30). Also to be seen, but more rarely, is amygdaloidal texture (figure 11). This results from the attempted escape of volatiles from the magma during flow, forming gas holes (vesicles) that were subsequently filled with other minerals (usually white calcite or black chlorite). Flow brecciation is the most common texture of all and occurs when the viscous surface crust of the lava solidifies whilst the interior of the flow is still moving. The crust is thus broken into angular blocks which are rafted along the surface (figures 12 and 13). These rocks are easily confused with some of the fragmental rocks formed by volcanic explosion (agglomerate, see below) and are best identified in the field from the fact that in flow breccias the matrix holding the fragments together tends to be harder than the fragments and thus stands out on weathering. The opposite is the case for agglomerate — it is the fragments which stand proud of the matrix (figures 15 and 65).

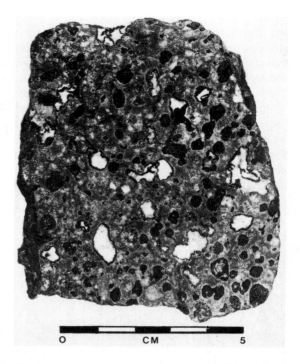

Figure 11   Amygdaloidal andesite, Duddon Valley. Volatiles rising towards the surface of the lava flow were prevented from escaping by the viscous semi-solid crust, and were trapped to form vesicles or gas bubbles. These holes were later filled by the minerals calcite (white) and chlorite (black)

Dacite and rhyolite lava

These rocks resemble each other in hand specimen and can be differentiated only by chemical analysis. They are richer in silica than andesite, more viscous and therefore tend to solidify more rapidly. Flow brecciation is therefore even more common than in andesite, with pressure from the interior, which is still liquid, breaking up the already solid crust. Flow banding in streaks of grey, green and pink, formed as the viscous magma moved, is also common and frequently displays flow folds (figure 14). Most dacites and rhyolites are pale grey, weathering to a white crust. In the eruptive cycle of the volcano they are related to ignimbrite, which is described below.

Figure 12 This diagram shows the characteristic mode of advance and the formation of flow-brecciated lava. The magma is viscous and advances slowly. The lava surface solidifies, but is fragmented and rafted forward by pressure from the liquid interior. The motion is rather like that of a caterpillar track on a tank. On final solidification the edge of such a flow is steep and rubbly

Figure 13 Flow-brecciated andesite lava, locality 7, Ulcat Row, Ullswater, excursion Da. The large angular fragments are softer than the matrix (the lava that flowed between the fragments), and they have become recessed on weathering. See figure 12 for an explanation of the mode of formation

Figure 14   Dacite (rhyolite). These rocks cannot be differentiated in the field. They are silica-rich and generally weather to a white crust, although freshly broken specimens may be quite dark and have a 'glassy' appearance. (A) Nodular (irregular lumps) flow-banded dacite from Little Hart Crag (excursion De). (B) Flow-banded dacite with folds formed during viscous flow

Figure 15 Coarse-grained tuff (agglomerate). (A) Volcanic breccia or agglomerate from Three Tarns, Bowfell (NY 249 060, excursion Fd). Note the angular fragments that stand out from the surface on weathering since they are harder than the fine-grained matrix. (B) Tilberthwaite tuffs, Hodge Close, Coniston (NY 316 017, excursion Gc). Large fragments of fine-grained tuff from a previous eruption are embedded in a fine-grained matrix. This rock is strongly cleaved (see figure 23) with the cleavage planes (C) crossing both fragments and matrix

PYROCLASTIC ROCKS

These rocks, formed by explosive action of the volcano, have the same range of chemical composition as the lavas. They are subdivided into: (1) accumulations of large blocks commonly within craters and on the adjacent slopes of the volcano (agglomerate and coarse tuff) since voicanic explosions are not usually powerful enough to eject fragments of this size very far from the vents; (2) fine ash (tuff) which can either also be deposited close to the volcano, drift for many miles in the upper atmosphere before it is deposited, or be deposited in water and reworked by currents; (3) water-saturated ash which descends valleys as hot torrential mudflows (lahars); (4) high-temperature blasts of incandescent ash, larger fragments and viscous magma (ignimbrite).

Agglomerate and coarse tuff (volcanic breccia)

Volcanic explosions, often a result of the sudden release of gas under pressure, shatter the rocks in and below the volcanic vent and can throw large blocks thousands of metres into the air. The larger fragments will fall back into the vent or on to the adjacent flanks of the volcano. Most fragments will have the composition of the lavas that have issued from the same volcanic centre and common rocks will therefore be basaltic, andesitic or dacitic agglomerate. It has already been indicated that there can be confusion between these rocks and flow-brecciated lavas, the difference in the field being that fragments in agglomerate and coarse tuff are harder than the finer-grained material in which they are embedded and project from weathered surfaces, whereas in flow breccias the fragments are recessed (figures 13 and 15).

Medium-grained and fine-grained tuff and bedded tuff

The finer ash has a much greater areal extent than the coarse material and will mantle the slopes of the volcano and be carried downwind for many miles. The recent eruption of Mount St. Helens resulted in substantial ash deposits up to 100 miles east of the volcano. The ash may fall on land (airfall tuff) or in water where it will be reworked by currents (volcaniclastic tuff), and in each case there may be alternating finer and coarser layers (bedded tuff); see figures 16, 17 and 18. The water-lain bedded tuffs of the Lake District are attractive, easily recognised rocks and are extensively quarried for ornamental stone which is sent to all parts of Britain and to other parts of the world. The bedding reveals the mechanisms of deposition clearly with many of the structures similar to those that can be seen in the recent sands of nearby Morecambe Bay. Ripples, channel structures and slump structures formed as the saturated ash slid down gentle underwater slopes, along with many other varieties of sedimentary structure (see figures 16, 17, 18, 31, 66, 81 and 82).

Volcanic mudflows (lahars)

It is frequently difficult to distinguish between mudflows and coarse tuffs since they are both composed of large fragments with a finer-grained matrix. The easiest way of distinguishing between them is by noting the varied nature of the fragments (pale and dark) that characterise the mudflows, which would have picked up anything in their path including rocks from previous eruptions. These rock fragments may therefore include basalt, andesite and dacite; the Mount St. Helens mudflows contained vast quantities of trees. This contrasts with the more uniform composition of coarse airfall tuff which largely consists of material from one volcanic centre. The Yewdale Breccia of Coniston is a formation that has been interpreted to be, in part, of mudflow origin (chapter 6, part G and figure 72).

Ignimbrite (ash flow) (Millward, 1979)

Ignimbrites are produced by the most violent and destructive of all volcanic eruptions. Their composition is usually rhyolitic to dacitic but there are some andesitic ignimbrites. The magma within the volcano is at a high temperature, silica-rich and viscous, building up high gas pressures. Eventually

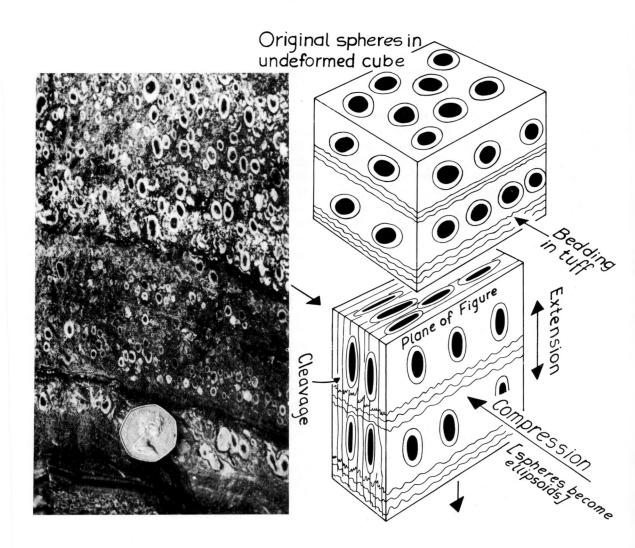

Figure 16 'Bird's eye' or accretionary lapilli tuff, Kentmere (NY 449 074). The origins are believed to have been as upper atmosphere accretions of fine ash into raindrops and hailstones. Witness the dense ash cloud associated with the 1980 eruption of Mount St. Helens. The hailstones then fall into water and the ash spheres melt out. They are generally finer grained on the edges. These particles (lapilli) became deformed during the Caledonian Orogeny (mountain building), when the high compressive stress flattened the rock in one direction and extended it in another as shown. Cleavage was developed during the compressive stage (see figures 22 and 23)

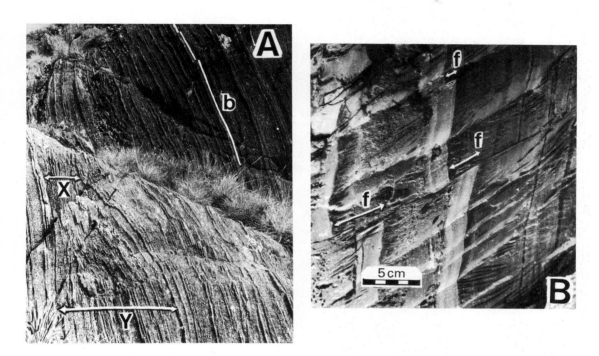

Figure 17 Yewdale bedded tuffs, Coniston (excursion Gb). (A) Formerly a near-horizontal ash fall deposit, now tilted to vertical by the Caledonian earth movements (see figures 21 and 22). This outcrop consists of alternating coarser and finer layers which are clearly revealed by differential weathering of the bedding (b). Note the variation in thickness of the deposit between X and Y, a common feature of tuffs. (B) A photograph of a gently sloping surface showing bedding, which is vertical and displaced by small faults (f). The latter are along the cleavage planes (see figures 22 and 23)

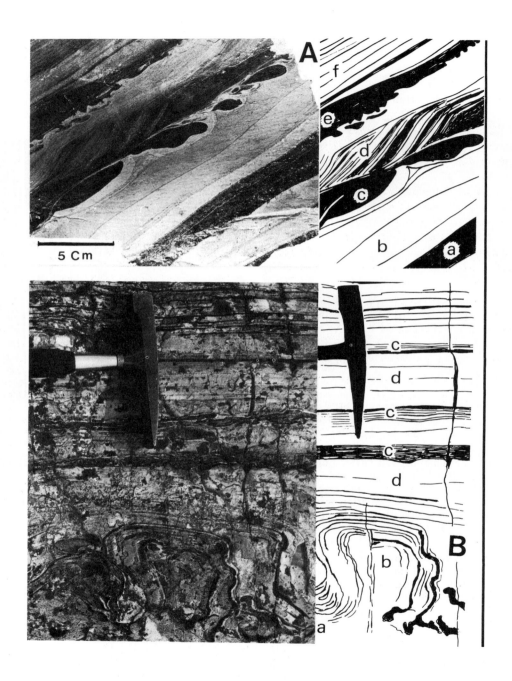

5 Cm

a minor event, such as a small earthquake, triggers the eruption and an incandescent mass of fragments, magma and gas explodes from the volcano usually travelling as a ground-hugging cloud (*nuée ardente*) down the side of the volcano and across adjacent hills and valleys often at more than a hundred miles per hour. Everything in the path of such an eruption is burnt and destroyed; witness the 1906 eruption of Mont Pelée on Martinique when virtually the whole 26 000 population of St. Pierre was killed, and the 1980 eruption of Mount St. Helens which flattened mature forest over an area of hundreds of square miles. Ignimbrites are common rocks in the Lake District and are generally easy to identify (see figure 19).

The molten constituents of the eruption are compressed into the shape of lenses by the weight of material deposited on top of them, and solidify as black glass. In rock specimens these appear as elongate black streaks in a pale coloured matrix. In the old Lake District rocks the original glass has been devitrified (that is, it has become microcrystalline) with the passage of time, and has been altered to chlorite and silica, but such features can be determined only by detailed microscopic analysis. The matrix, the fine-grained particles between the dark lensoid streaks and angular fragments (figure 20), consists of smaller rock fragments, ash and glass shards (broken pieces of glass) although a microscope is also necessary to see the latter. Characteristically, weathering results in a white surface across which the black streaks (fiamme) are easily seen (figures 19 and 20).

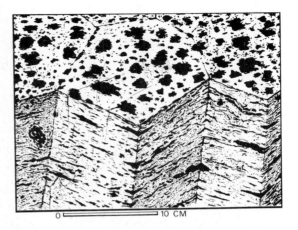

Figure 19 A diagram to show the characteristic features of an ignimbrite. High-temperature blobs of viscous magma and some solid fragments are deposited during a *nuée ardente* type of eruption. Much of this material can behave as a liquid (magma) and is liable to flow, hence the name 'ash flow tuff'. However, the predominant texture of the rock is a result of the weight of material above, which flattens the blobs into disc shapes before they become solid. These disc shapes are called fiamee and the texture is described as eutaxitic. The already solid fragments retain their angular shape and the fiamme are compressed around them. Should the ignimbrite be subjected to appreciable flow the fiamme become stretched out and thinner as in figure 20(B). Columnar jointing shown here (see also figure 78) is another common feature of ignimbrites and develops when hot materials, ash flows or magmas cool and contract

Figure 18 (opposite) Bedded volcaniclastic tuff, formerly volcanic ash deposited in water then drifted and reworked by currents and waves. The resulting structures resemble those of ordinary river, lake and sea sediments (sand and silt) subjected to the same processes. (A) Ornamental tuff from Hodge Close, Coniston (excursion Gc): a, medium-grained dark tuff; b, fine-grained bedded tuff; c, medium-grained tuff with 'load structures' formed by 'new' ash deposits settling unevenly on the water-saturated fine ash below (asymmetry of these structures is a result of some current flow); d, cross-bedded unit formed by small-scale currents or wave ripples; e, disturbed bedding (load structures) caused by deposition on water-saturated ash, and subjected to some current agitation; f, regular bedding resulting from steady ash fall of coarser and finer material. (B) Bedded tuff (Seathwaite Fells Tuff, Bowfell, excursion Fd). Water-lain tuffs that have been subjected to wet sediment slumping down gentle slopes: a, coarse tuff with moderately large fragments; b, a fold formed in consolidated saturated ash by slumping (convolute bedding); c/d, alternations of fine-grained and medium-grained tuff with regular bedding, indicating fluctuating but near-continuous eruption

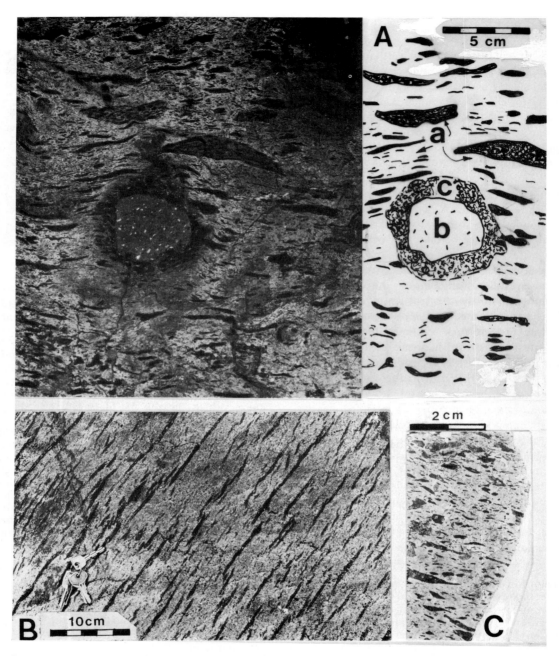

Figure 20 Characteristic ignimbrite (see figure 19). (A) From White Crag, Langdale (excursion Fa) a well-defined streaky (eutaxitic) texture is present. The dark streaks (a) (fiamme), originally semi-liquid blobs, were flattened to discs before solidifying by the weight of material above. They are embedded in a fine-grained ash (tuff) matrix. (b) is a fragment that was already solid, but very hot, at the time of eruption. It reacted chemically with the surrounding hot ash (c) which now appears dark. (B) From Knock Pike in the Cross Fell Inlier, east of the Lake District (NY 69 29), it can be seen how the fiamme were pulled out into long thin streaks by flow (ash flow); this is referred to as a parataxitic texture. The ignimbrite of Knock Pike has been tilted to an angle of 60° by the Caledonian earth movements. (C) Ignimbrite from Tertiary volcanics of South Arabia. This type of rock occurs world-wide

4

The Caledonian Orogeny

I observed in the introduction that the Lower Palaeozoic epoch in Britain culminated in 'continental collision' at the end of the Silurian Period, when the ancient American and European continents drifted together to form one southern-hemisphere Eur-American continent (figures 3 and 6). The resulting compression raised the Caledonian mountains, a range of Alpine proportions that extended from Norway to the North American Appalachians; it also folded and faulted the rocks along the margins of both former continents so that complex structures are now to be seen from northern Scotland to South Wales. These structures vary from region to region and in the Lake District an important influence was the rock succession of the soft, easily deformed Skiddaw Slates, followed by the resistant and hard Borrowdale Volcanics and then by the weaker Silurian Slates and Sandstones (figures 1 and 2). The structures in these rocks can be considered under three main headings: folding, cleavage and faulting, the latter including the formation of joints and mineral veins.

FOLDING (figures 21 and 22)

The Skiddaw Slates (mostly mudstones) were crumpled into complex small-scale folds often with amplitudes of less than a metre. A walk along the Caldew Valley just below Carrock Mine (NY 330 325) reveals structures of this kind, which have been described in detail by Roberts (1971) and are perfectly exposed in the stream bed. It has to be said, however, that some of these folds have disputed origins; the alternatives are (1) folds produced by the slumping and sliding of wet sediment down gentle slopes, shortly after deposition, and (2) folds produced by the compressive stresses of mountain-building earth movements at a much later time. Both processes were effective in the Skiddaw Slates, and a comparison of figures 21, 24 and 65 will demonstrate that an unequivocal solution is evasive. The Silurian Bannisdale Slates above (and younger than) the volcanics were also crumpled into small folds, well seen alongside the A6 just south of Shap summit and illustrated in the Yorkshire Geological Society's book *The Geology of the Lake District* (Moseley, 1978). They are also well seen in the Coniston Grits alongside the M6/A685, 1 mile south of Tebay (Moseley, 1972), and in both cases are undeniably of tectonic origin. The volcanic rocks however are much stronger than the Skiddaw and Silurian (Bannisdale) Slates and, although folded, they were not crumpled on such a small scale. Indeed the amplitudes of folds in the volcanics are to be measured in kilometres rather than metres and most views of distance fellsides will show strata inclined in one direction only, representing single limbs of the larger folds (figures 17 and 23).

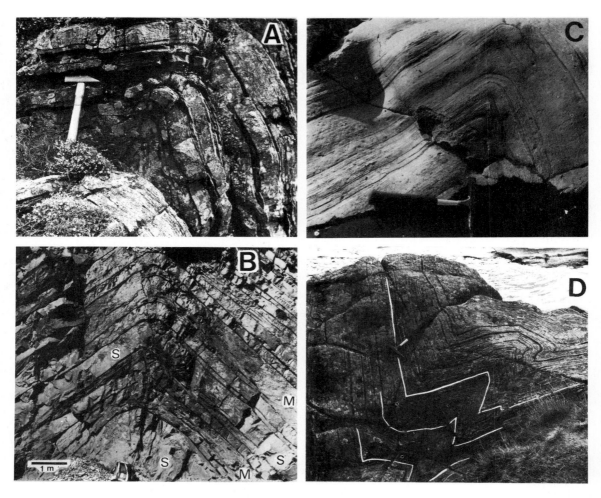

Figure 21 Characteristic small-scale folds from the Lake District. (A) An anticline in Loweswater Flags, Skiddaw Group, Whiteside (NY 167 220). The rocks are sandstones (greywackes). (B) An anticline in beds transitional from Coniston Grit to Bannisdale State, Silurian, Shap Fell. The locality is alongside the A6 road at Shap summit (NY 555 055). The rocks are alternations of greywacke-type sandstone (S) and cleaved mud-stone (M). (A) and (B) were formed by compressive (tectonic) forces during the Caledonian earth movements. (C) and (D): Folded Skiddaw hornfels (Skiddaw Slate that has been thermally metamorphosed and recrystal-lised by heat from the Skiddaw Granite intrusion — see also figure 24), Caldew Valley NY 330 226. This is now a very hard rock and although the cleavage is no longer visible the folded bedding is easily seen. The question is whether these folds resulted from tectonic forces, as in the cases of (A) and (B), or were formed by the slumping of wet sediment shortly after deposition. I personally favour a tectonic hypothesis because of their angular nature and constant orientation (Roberts, 1971), but compare with figures 29C, 30 and 65A

CLEAVAGE

Cleavage results from the high stress imposed on rocks during the lateral compression of orogeny (mountain building). Certain minerals, especially clay minerals and mica, recrystallise under the impact of this stress and tend to grow in a direction at right angles to it. The result is a rock that no longer splits along the original planes of deposition (bedding), but along this new direction (cleavage) (figures 14, 16, 22 and 23). The effect is greatest on the finer-grained rocks (those with the smallest particles), and those with an abundance of flaky minerals such as mica, chlorite and clay minerals. Consequently the former muds of the Skiddaw and Bannisdale Slates are frequently strongly cleaved, as are the fine-grained chlorite-rich tuffs of the volcanics, but the massive hard sandstones (Loweswater Flags, interbedded with Skiddaw Slates, and Coniston Grits in the Silurian), and the massive lavas of the volcanic sequences, show only weak cleavage. The ornamental green slates of the Lake District, which were formed as chlorite-rich bedded tuffs, owe their economic importance to a good cleavage that makes for easy splitting across the depositional or bedding layers, revealing attractive patterns of greys and grey-greens.

FAULTING

The slow but irresistable movement of continents during collision also results in fractures in Earth's crust (faults) as one segment slides past another. This movement is not continuous; it occurs in sudden jerks following a steady build up of stress when the rocks reach breaking point. The resulting fracture is experienced as an earthquake which can displace surface features (roads, fences, water courses etc.) by as much as 30 metres and reduce towns to rubble. This is of serious concern to countries such as Yugoslavia, Turkey and Iran which experience the earthquakes caused by continental collision of the present day. A single phase of continental collision is short in relation to geological time but it is eternity in terms of the human life span. It is likely to last more than 10 million years and consequently one earthquake every 100 years along one fault can result in a substantial displacement; for example the San Andreas Fault of California and the Alpine Fault of New Zealand each have displacements exceeding 100 miles.

In the Lake District ancient faults are commonplace; displacements, estimated from the rocks that outcrop on opposite sides of the fault, range from a few feet to over a mile. For example, the Coniston

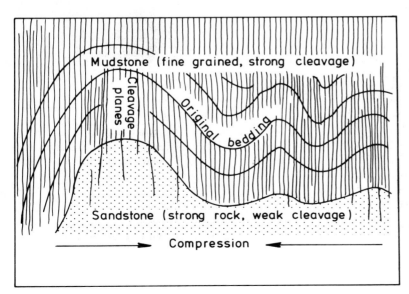

Figure 22 Structures formed during the Caledonian Orogeny (mountain building). Bedding planes that were originally horizontal are folded into anticlines (up folds) and synclines (down folds). Intense compressive stress results in recrystallisation, especially of micas and clay minerals, so that they become aligned at right angles to the stress. In consequence, the rock now has a new splitting direction (cleavage), which is more pronounced in mudstone (and fine ash) than in sandstone (and massive lava)

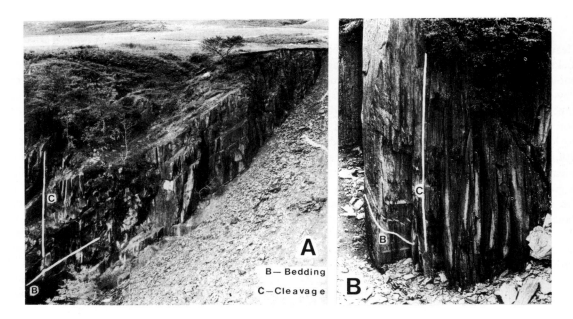

Figure 23   Relation between bedding and cleavage. These rocks were folded during the Caledonian Orogeny (earth movements), and the originally near-horizontal bedding (depositional layers) was tilted to angles varying from a few degrees to vertical. The compression resulted in recrystallisation of certain minerals (for example, micas and clay minerals) so that they became aligned at right angles to the stress direction. The rock now splits in this new direction (cleavage) rather than along the original bedding planes. (A) Banishead Quarry, Coniston (SD 280 960) showing the Brathay Flags inclined about 35° south-east with near-vertical cleavage (excursion Ga). (B) Tilberthwaite tuffs near Coniston (NY 305 008) showing a dominant vertical cleavage cutting obliquely across occasional bedding traces

Limestone is displaced 1 mile by a major north-south fault that runs through Thirlmere and across Dunmail Raise, and is then followed by the main road (A593) and subsequently by Coniston Water (figure 26). Most of the faults are much smaller than this, with displacements of less than 100 metres, but they are still important structures as are the joints, subparallel to the faults, and the related mineral veins. Most of the joints are formed by a similar process to that which formed the faults; they are generally less than a metre apart and are distinctive features of all outcrops but unlike the faults they do not displace the rocks. Mineralising solutions frequently followed these planes of weakness, resulting in veins rich in copper, lead and other minerals that have been so important to the Lake District economy in historical times (Shackleton, 1968; Holland, 1981). In the field a principal characteristic of most of these fractures is that they form nearly vertical zones of weakened or shattered rock easily picked out by erosion. They give rise to straight gullies, easily seen on distant fellsides and revealed to perfection on aerial photographs. Examples of faults, joints and veins are to be seen on many of the illustrations in this book (see particularly figures 33, 48, 54, 55, 56, 57, 62, 63, 72 and 79).

Igneous Intrusions

From early in the Ordovician until the end of the Silurian the continental margin of ancient Europe was subjected to a great deal of underground igneous activity. Magma forced its way through other rocks towards the surface as igneous intrusions which are now exposed at the surface as outcrops of granite, gabbro, diorite and dolerite (figures 25, 26 and 34) (Firman, 1978). The intrusive activity can be subdivided into two main phases, the first related to the Ordovician volcanicity and the second to the continental collision and mountain building at the end of the Silurian Period.

It has been explained in Chapters 1 and 2 that the Ordovician volcanoes would have been supplied with magma from below. Magma chambers would have existed within the granitic crust (figure 5) and from these the magma would have risen along pipes and fissures (plugs and dykes) to feed the volcanic centres. There are large and small intrusions that may be attributed to this phase of activity. The gabbro and related rocks of Carrock Fell, for example, may well have been a magma source for some of the Eycott volcanic rocks in the north. The Ennerdale Granophyre and Eskdale Granite, both intrusions of several miles diameter, are contemporaneous in age, or not much later than the outpouring of the Borrowdale Volcano, and originally may have been magma chambers for some of the acid volcanism (dacite and ignimbrite). Another likely source of some of

the andesitic and basaltic rocks is the Haweswater Complex of dolerite and gabbro, whilst the Upper Ordovician St. John's microgranite may have been a source for some of the latest dacites and rhyolites (figure 26, 27, 36 and 41).

The Silurian Period was almost free of igneous activity, but the end of the Silurian and the early Devonian saw the emplacement of a hugh batholith beneath the Lake District and the northern Pennines. This is an intrusion principally of granitic rocks which reaches the surface in only two comparatively small areas as the Skiddaw and Shap Granites (figures 25 and 26), but its presence underground has been demonstrated geophysically (by gravity surveys) and by one bore hole in the northern Pennines.

The heat from all these intrusions has resulted in alteration (thermal metamorphism) of the rocks that they penetrate. For example, the heat from the Skiddaw Granite recrystallised much of the Skiddaw Slate adjacent to the contact, resulting in the rock variety known as hornfels. This is well exposed along the River Caldew west of Mungrisedale. Even where the heating was less intense, new crystals were able to grow in the slates; for example, the andalusite (chiastolite) which has grown across both bedding and cleavage (figure 24) has resulted in the well-known chiastolite slate, to be found in University teaching collections all over Britain.

Figure 24 Thermally metamorphosed Skiddaw Slate (mudstone) from Glenderaterra (NY 300 270). The sedimentation layers (bedding B) consist of alternations of fine-grained dark clay and slightly coarser-grained paler silt. They are highly inclined and have been contorted by folding. The white crystals (A) are andalusite and grew randomly at all angles to the bedding in response to the heat (thermal metamorphism) from the underlying Skiddaw Granite (figure 1)

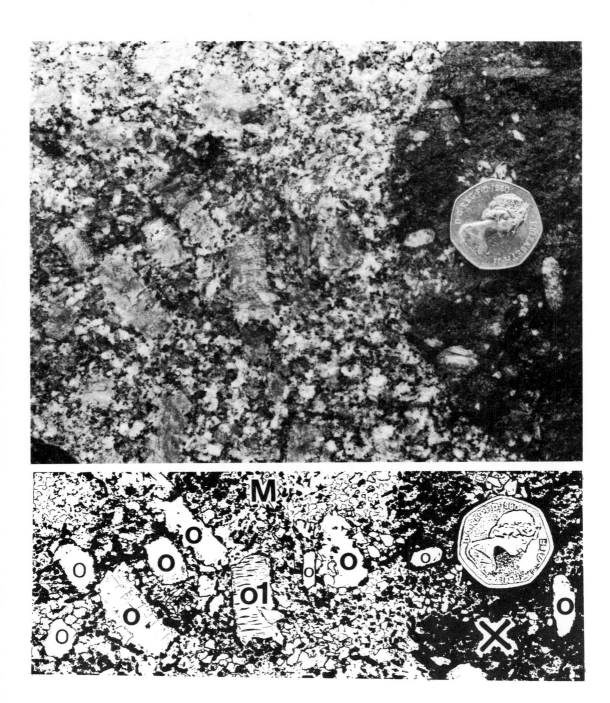

Figure 25 Shap Granite (NY 557 083) — the surface expression of a large early Devonian intrusion (batholith) that underlies the Lake District and north Pennines (see figure 1). It is a well-known ornamental stone and can be seen decorating shop fronts all over Britain. The rock has a coarse grain with large crystals of (pink) orthoclase feldspar (O). Close inspection of these crystals will reveal a patchy striping (O1), known as perthitic texture. The matrix (M) is mostly made up of smaller crystals of feldspar and quartz (pale) and biotite mica (dark). The dark area to the right (X) is part of a block of the surrounding volcanics which was incorporated into the granite magma and partly remelted (xenolith). Note the feldspar crystals that have grown in this xenolith.

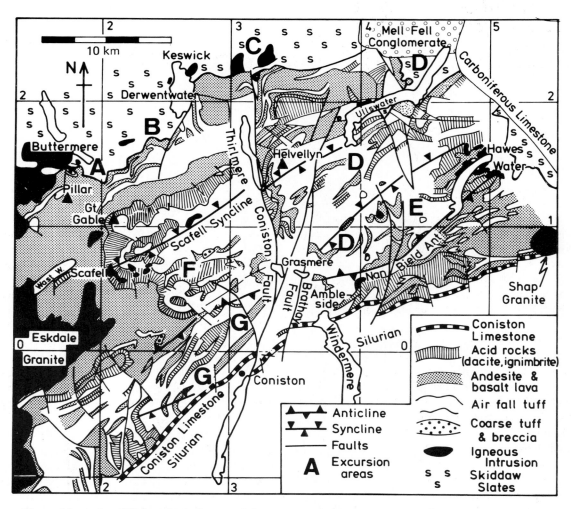

Figure 26   A simplified geological map of the Borrowdale volcanic outcrop of the central fells. The locations of the excursions are indicated

6

Field Excursion Itineraries

The excursions (see figure 26) cover the whole width of the Borrowdale volcanic outcrop, and embrace part of the older Skiddaw Slates in the north and the younger Coniston Limestone in the south. They are of varying geological difficulty but I believe both those new to geology and experts will find them rewarding. The former should not be discouraged by the difficulties and more advanced treatment of some of the excursions, particularly A, B and G; there is plenty to be seen on a generalised level. Nor should the latter be put off by what may seem to be a 'simple' treatment. By following the routes indicated it will be possible to observe many details not mentioned in this text. Other general references and field guides to the Lake District are Mitchell (1956, 1970), Shackleton (1975), Moseley (1978) and Cumberland Geological Society (1982).

The maps given in this book show the general route of each excursion. It is recommended that these be supplemented by the appropriate walking maps for the area. The whole of the Lake District is covered by an Ordnance Survey 1:50 000 (about $1\frac{1}{4}$ inches to 1 mile) Tourist Map. Even better are the 1:25 000 (about $2\frac{1}{2}$ inches to 1 mile) Ordnance Survey series. The relevant maps are noted at the beginning of each excursion. Four 'Outdoor Leisure' maps are also available for the Lake District at a scale of 1:25 000. All these maps are obtainable from your local Ordnance Survey stockist.

A. BUTTERMERE TO HONISTER (1:25 000 maps NY11, NY21)

This is an area of great interest (figure 27) but much of the geology is exceedingly complicated and it has therefore been necessary for the excursion comments to be graded according to the geological knowledge of those concerned. I have found this a difficult task but I have attempted to describe each locality first in terms that will be understood by the beginner and then to expand with comments more suitable for those with experience. The rocks involved are Skiddaw Slates, Borrowdale Volcanics and a number of igneous intrusions. Figure 27 is a geological map of the area and figure 28 shows a view of the Buttermere/High Stile range from the viewpoint indicated on figure 27.

The Skiddaw Slates (or Skiddaw Group), well exposed around Buttermere, are the oldest of the Lake District rocks, and are overlain by the Borrowdale Volcanics. The tectonic structures of the slates, the variety of the volcanic rocks and intrusions and the nature of the junction between the slates and volcanics form the essence of the excursion notes below. Some of the details are particularly complicated and beyond the scope of this book, but for those wishing to pursue the problems further there are a number of more specialised publications that should be consulted (Ward, 1876; Clark, 1964; Simpson, 1967,

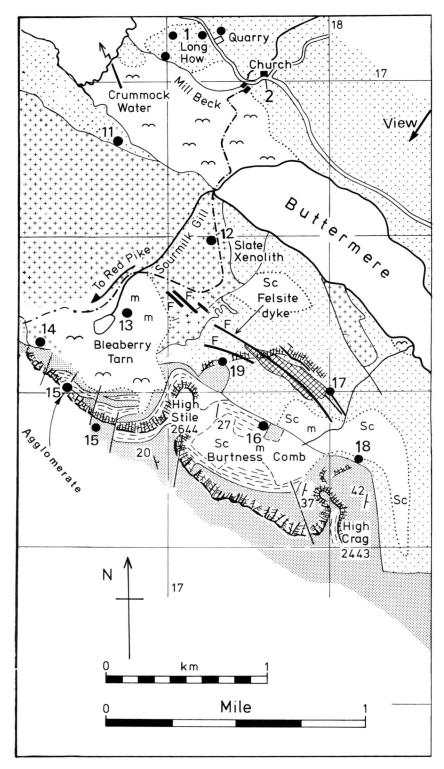

Figure 27 An outline geological map of the Buttermere — Honister area.

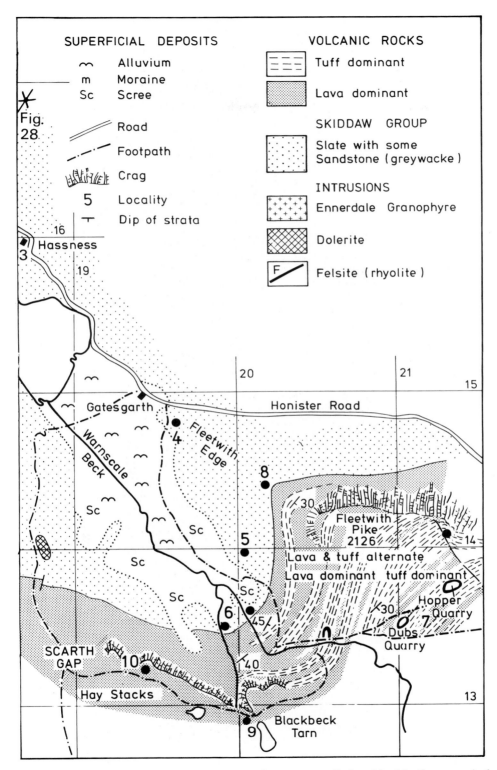

SUPERFICIAL DEPOSITS

- ⌒ Alluvium
- m Moraine
- Sc Scree

- Road
- Footpath
- Crag
- 5 Locality
- Dip of strata

VOLCANIC ROCKS

- Tuff dominant
- Lava dominant

SKIDDAW GROUP

- Slate with some Sandstone (greywacke)

INTRUSIONS

- Ennerdale Granophyre
- Dolerite
- F Felsite (rhyolite)

Fig. 28

16
Hassness
3
19

20
21
15

Gatesgarth
Honister Road

Warnscale Beck

Fleetwith Edge
4

8

Sc

Sc
5

Sc

30
Fleetwith Pike 2126

14
Lava & tuff alternate
Lava dominant tuff dominant

Sc

Sc
6
45
30
Hopper Quarry
7
Dubs Quarry

SCARTH GAP
10
Hay Stacks
40

Blackbeck Tarn
9

13

Interesting localities are shown but they need not necessarily be visited in the numerical order indicated

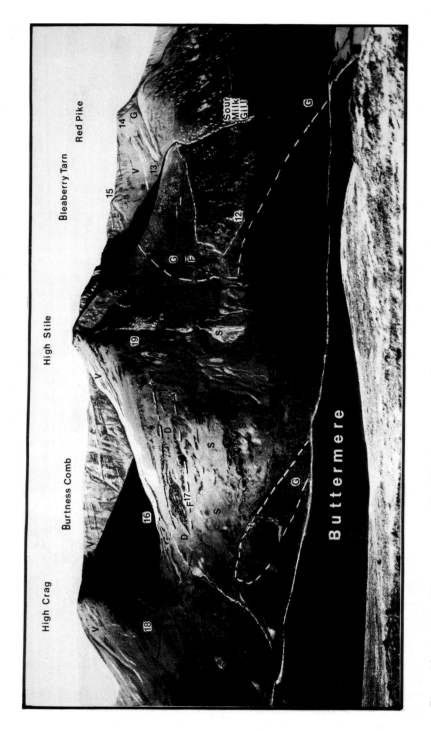

Figure 28   A view of Buttermere and the High Stile range from High Snockrigg (NY 187 169). The localities mentioned in the text are indicated. G — Ennerdale Granophyre, D — Burtness Dolerite, F — felsite dykes, V — Borrowdale Volcanics, S — Skiddaw Slates. The principal boundaries between these rocks are shown by dashed lines

Shaw, 1970; Soper, 1970; Jeans, 1971, 1972; Moseley, 1972, 1975, 1978, 1981; Jackson, 1978; Soper and Moseley, 1978; Wadge, 1978). The Skiddaw Slate structures and the nature of the junction between the slates and the volcanics are perhaps the most difficult problems. For example the original muds of the Skiddaw Slates were first contorted by the slumping and sliding of wet sediment down gentle submarine slopes, producing fold and fault structures that are almost identical to those produced by mountain-building (tectonic) processes (see also figures 21, 66 and 81). Subsequently, tectonic structures of similar style were superimposed on these sedimentary structures; indeed this happened on several occasions, first prior to the formation of the Borrowdale Volcanics, when north-trending folds were formed, and second and third during the main Caledonian Orogeny at the end of the Silurian Period, when NE-trending folds and cleavage were formed. The complication and difficulties of interpretating all the resulting structures can well be imagined.

The second problem, the nature of the junction between the slates and the volcanics, is just as difficult to solve unequivocally. The alternatives are (i) whether deposition of Skiddaw sediment was followed without a break by the onset of volcanism, (ii) whether there was an appreciable time gap between the two processes during which the Skiddaw sediments were folded and eroded or (iii) whether the Skiddaw Group was followed by a major orogeny and subjected to severe folding and cleavage before the volcanicity started. Whatever the interpretation, the Skiddaw Slates have been strongly deformed and small-scale folds will be seen at most exposures. The overlying volcanics include basalt, basaltic andesite and andesite lavas, and bedded tuffs such as those worked in the Honister quarries. They are mostly strong rocks that resisted the crumpling of the Caledonian earth movements, although the tuffs do have a strong cleavage (figure 29A). Indeed this is the important economic factor as far as the quarrying of the tuffs is concerned, since the rock can be split into thin layers along the cleavage.

In addition to the slates and volcanics there are a number of igneous intrusions, the largest of which is the Ennerdale Granophyre. The latter is now known to be of Ordovician age (Rundle, 1979); it is essentially a medium-grained pink microgranite with small crystals, mostly quartz and feldspar, that can be seen with a hand lens. There is also a prominent basic intrusion (altered dolerite) – a dark, medium-grained to coarse-grained rock of approximately the same composition as some of the lavas – and there are dykes of rhyolitic to dacitic composition, usually referred to by the generalised field term of felsite because of identification difficulties. The dykes are vertical sheets of igneous rocks only a few feet wide which originally rose through cracks in the Skiddaw Slates.

The excursion numbers on figure 27 can be followed in any order. It will depend whether the object is to study slate outcrops near Buttermere, by far the most difficult geologically, or to walk across Fleetwith Pike, Haystacks and High Stile.

1. 172 172, Long How. Thirty metres beyond the gate, nearly opposite the quarry, there are excellent exposures of small-scale folds in Skiddaw Slates. The folds are clearly revealed by pale silty layers interbedded with mudstones. The outline is easy to appreciate but details of the folds are for the more advanced student. They do not represent a simple crumpling of the strata but are steeply plunging and have had a polyphase history; that is, these rocks have been subjected to more than one phase of folding and in fact two phases referred to as F1 and F2 can be demonstrated at this locality (Soper and Moseley, 1978). Further into Long How at 170 171, above the path alongside Mill Beck, there is another example of polyphase folding, in this case involving the second and third phases (F2 and F3), with F3 open recumbent folds with horizontal axial planes (Soper and Moseley, 1978; Moseley, 1972). It is also worth visiting Buttermere Quarry (173 172) where structures are much more complex than they appear at first sight. Load casts on a sandstone at the top of the quarry show that the highest part of this vertical sequence is towards the south-east (see figure 18A for the significance

of these structures when determining top and bottom (way up) of strata).

2. 175 170, Buttermere Church. Careful examination of the top surface of the slate outcrop adjacent to the church will reveal complex minor folds on less than a metre scale. This outcrop in fact exhibits three phases of deformation. The first two phases are represented by steeply plunging Z folds with subaxial plane cleavage and the third phase by gentle recumbent folds with horizontal cleavage.

3. 186 159, Hassness. This locality has been described by B. C. Webb. It is similar in many respects to the Long How outcrop. Examine the horizontal surface of a prominent *roche moutonnée* (a glacially smoothed outcrop) and obvious small folds will be easily seen. The details are much more complex and those interested are recommended to consult Webb's paper (Webb, 1972).

4. 196 148. Localities 4–10 form an excursion route quite different to the localities just described. Starting at Gatesgarth there are two footpaths. The first begins the steep ascent of Fleetwith Edge and the other skirts the fellside to Warnscale Beck. Locality 4 at the bottom of the Fleetwith Edge consists of greywacke sandstones with interbedded mudstones (slates) of the Loweswater Flags Formation.

5. 200 140. Follow the footpath to Warnscale Beck. There are a number of small Skiddaw Slate crags on the steep slopes descending from Fleetwith Edge, most of which show small-scale folds. The folding at the above grid reference, which is about 200 feet above the path, is particularly good with tight F2 folds and small shears.

6. 201 136. The junction between the Skiddaw Slates and Borrowdale Volcanics is exposed both in Warnscale Beck (201 135) and Black Beck (199 135) (Moseley, 1975; Soper and Moseley, 1978). In Warnscale Beck the Skiddaw Slates below the junction are strongly folded. This is best seen in the stream bed where the slates have been polished by water flow, and the pale silty bands contrast strongly with the dark mudstones. The actual junction is a sharp plane inclined 60° south-east, and is a fault. The volcanics upstream of the fault are flow-jointed basaltic andesite lavas that dip

uniformly towards the south-east. In Black Beck there is a conglomerate at the junction (subspherical boulders of mudstone and volcanics) which seems to indicate a possible unconformity (Moseley, 1975; Soper and Moseley, 1978).

7. 213 137, Hopper Quarry. Take the most northerly footpath which climbs steeply above the junction locality. It crosses alternations of lava and tuff, eventually reaching the prominent tuff bed worked in Hopper Quarry (permission to visit the quarry should be requested). There are in fact two prominent tuff beds worked in the Honister quarries, the lower known as the Honister 'vein' (a quarryman's term) and the upper as the Hopper 'vein'. Both are bedded volcaniclastic tuff much valued for their ornamental properties (figures 17, 18 and 29A). Good specimens can be picked up on the spoil heaps but much better ones can be purchased cheaply at the main works on Honister Pass. I hesitate to refer to details of the Hopper Quarry since what may be seen one year is likely to have been quarried away the next.

From this locality several routes are possible. One may prefer to walk over Fleetwith Pike, returning to Gatesgarth down Fleetwith Edge. Alternatively there is a most interesting route across Haystacks, returning to Gatescarth via Scarth Gap.

8. 202 144, Fleetwith Edge. The slate–volcanic junction is exposed at several points across the ridge. Both north and south of the ridge conglomerates similar to that of locality 6 can be seen (Soper and Moseley, 1978).

9 and 10. 201 129 to 194 132. A well-marked footpath traverses Haystacks and the exposures described here are perfectly obvious without leaving the path. Of particular interest is the impressively flow-folded and flow-jointed andesite lava well exposed between Blackbeck Tarn and Haystacks summit (figure 30). Similar folds will form in most viscous fluids subject to flow, but although they resemble folds formed by tectonic processes (earth movements) the mechanisms are different.

The localities from 11 to 19 cover a different area to those just described and I would expect

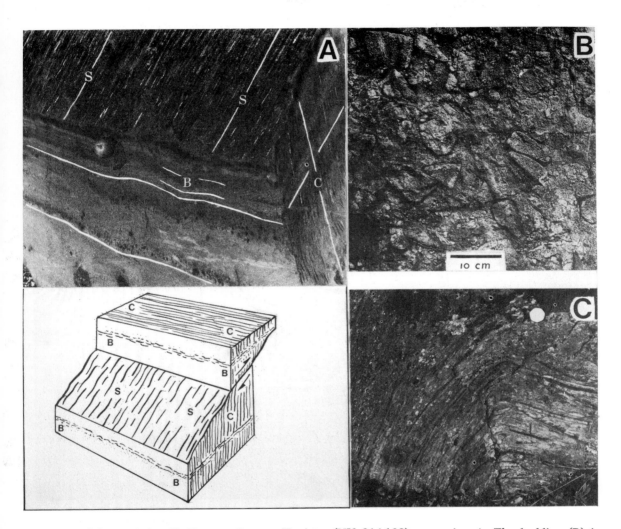

Figure 29 (A) Bedded tuff, Hopper Quarry, Honister (NY 214 138), excursion A. The bedding (B) is revealed by finer and coarser-grained layers. Cleavage (C) is strong and facilitates the quarrying of the rock; it is the plane of easy parting. There is also an oblique cross-cutting surface that is quartz-veined and heavily grooved. This is a slickensided plane (S) and represents a small thrust fault with the relative rock movement as shown. Other similar planes along which there has been no movement are often to be seen; these are master joints. Request permission to visit this quarry and remember that in a working quarry the exposures continuously change. (B) Agglomerate or volcanic breccia, locality 15, High Stile, excursion A. This outcrop is believed to be part of a former volcanic vent. (C) Flow banding in a felsite dyke, locality 17, excursion A

Figure 30 Flow folds in andesite lava on Haystacks (NY 194 132, locality 10). In this case 'flow joints' (analogous to bedding planes of a sandstone) are contorted into tight folds and were formed as the lava advanced in flow, not by the much later stresses of the Caledonian mountain building

them to form the nucleus of another excursion. Starting once again at Buttermere Village, take the footpath leading south towards Bleaberry Tarn and Red Pike. Along some parts of this route there is a wealth of outcrop and some of the locations, which were first demonstrated to me by Don Aldiss and Kevin Smith when they were students at Birmingham University, may be difficult to find.

11. 167 166, near Scale Bridge. The junction between Skiddaw Slates and granophyre can be seen hereabouts. Pinkish microgranite is in contact with pale-grey baked Skiddaw Slates. Within 10 metres the slate has become its normal dark colour and it appears that thermal metamorphism (alteration by heat from the intrusion) was not very great.

12. 173 159. Immediately east of the Bleaberry Tarn footpath normal pink microgranite contains xenoliths of Skiddaw Slate. These are blocks broken from the Skiddaw Slates during intrusion and enclosed in the microgranite (granophyre). There are also microgranite veins (narrow bands) within the slate blocks.

13. 166 154, Bleaberry Tarn. This hollow was eroded in recent times during the last glaciation of the region. It represents the site of a former corrie glacier that merged with the main Buttermere glacier at glacial maximum. The hollow now hangs 400 metres above the Buttermere valley. The debris from the corrie glacier (moraine) now surrounds the tarn and forms irregular hummocks that conceal the solid rock.

14. 162 153, Red Pike. Close together on the eastern face there are vertical contacts between Skiddaw Slates and Borrowdale Volcanics, and between Skiddaw Slates and Ennerdale Granophyre.

15. 163 151 to 168 148, Chapel Crags. The upper part of the north-facing escarpment between Red Pike and High Stile exhibits a wealth of volcanic features. At 163 151 there is a prominent outcrop of agglomerate (figure 29B). It consists of an accumulation of large volcanic blocks and probably represents the site of a former volcanic vent. A little further south-east there are well-bedded tuffs with a variety of sedimentary structures (figure 31). There are also interbedded andesite lavas, in one case with a strongly transgressive base cutting into tuff.

16. 176 148, Burtness Comb. Agglomerate with andesite fragments up to 1 metre in diameter can be seen. These outcrops were described by Clark (1964) as the site of a former vent.

17. 180 150, Burtness dolerite and felsite. The dolerite is intruded into Skiddaw Slates, forming several parallel sheets, and is about 60 metres thick in all. These sheets are steeply inclined and appear to have been members of a multiple intrusion (intruded one after the other). The rock is medium-grained to fine-grained but dark (altered) pyroxene and pale feldspar crystals can be seen with a hand lens.

At the base of the dolerite crags there are good exposures of felsite (probably dacite) dykes. These are steeply inclined sheets of pale rock up to 5 metres thick, which cut across the dolerite and are therefore younger. The rock is strongly spherulitic (small spheres formed by the devitrification of a former glassy rock). The spherulites are arranged in bands and reveal complex flow folding. A very small path leads from here, across the crags towards Bleaberry Tarn (figure 29C).

18 and 19. 182 146 and 173 152, High Stile to High Crag. These localities exhibit the controversial junction between the Skiddaw Slates and the Borrowdale Volcanics (see above and the Borrowdale excursion B). North of High Stile (location 18) the junction is marked by a thin conglomerate of volcanic pebbles, which rests on cleaved Skiddaw mudstone. Above it there is andesite lava and tuff. At the northern end of High Crag (location 19) several metres of conglomerate composed of mudstone and andesite pebbles mark the junction. It rests on weathered Skiddaw Slate and is followed by steeply dipping andesite lava. My opinion is that both these localities represent relatively small unconformities rather like those described for excursion B.

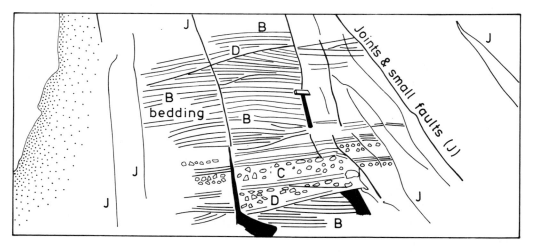

Figure 31 Bedded tuff on High Stile (NY 165 148, locality 15). The bedding (B) is well seen in coarse (C) and finer layers. There are minor erosional surfaces (disconformities) where fresh deposits of tuff cut across earlier ones (D)

B. BORROWDALE, HIGH SPY AND NEWLANDS (1 : 25 000 map NY 21)

In many respects this excursion resembles the Buttermere excursion and is concerned partly with the Skiddaw Slates, partly with the volcanics and partly with the junction between them. Many of the problems are similar to those outlined for the Buttermere area, and the geology likewise varies from that which can be easily understood by those with little geological experience to details that are difficult even for experts. I will try to indicate which of the localities are complex in the locality descriptions.

1. 252 167, Quayfoot Quarry (figure 32). Bedded andesitic tuff with steep ENE cleavage.
2. 253 175, Grange. Skiddaw Slates are exposed on a glacially smoothed slab (*roche moutonnée*) adjacent to the river. The bedding is tightly folded, rather like the Buttermere Church exposure; the trend is north-south and there is no obvious cleavage. The details of this exposure are not easily understood.
3. 246 176, Greenup Sike. This outcrop on the stream bank is for the advanced student; it shows a fold in Skiddaw Slates with a vertical plunge. This is brought out by a thin sandstone band in slate, seen only from one vantage point.
4. 249 170, Scarbrow Wood. This is a classical exposure of the junction between Skiddaw Slates and Borrowdale Volcanics, and was first described over 100 years ago. It has always been controversial and remains so. Several alternative explanations have been proposed by different research workers: (a) the two groups are conformable with virtually no time gap between them; (b) there is a small unconformity, the Skiddaw Slates being subjected to minor folding, uplift and erosion before deposition of the volcanics; (c) a major unconformity exists — that is, a major period of folding with imposition of cleavage before volcanism started (see Simpson, 1967; Soper, 1970; Jeans, 1971, 1972; Mitchell *et al.*, 1972; Moseley, 1972, 1975, 1978; Soper and Moseley, 1978; Wadge, 1978).
5. 247 170. The fellside immediately above Scarbrow Wood is easier to inspect, not being overshadowed by trees. The problems are similar to those of locality 4, however, and equally difficult to solve.
6. 245 169. A little farther up the hill, gently dipping andesitic tuffs are separated from Skiddaw Slates by a vertical fault.
7 and 8. 242 170 to 241 171. Below Blea Crag there are two similar exposures at the base of the volcanics, first described by Soper (1970). They consist of conglomerate with subspherical boulders of (Skiddaw) mudstone and volcanics, which rests on Skiddaw Slate (mudstone).
9. 243 164, Goat Crag. Follow the footpath that passes west of Goat Crag (figure 32). This crag is formed of andesite lava dipping 25° to the south.
10. 235 162, High Spy. This region consists of alternating andesitic lavas and bedded tuffs (figures 32 and 33). The westward view to Hindscarth (figure 35) clearly shows the Squatt Knotts dolerite-diorite intrusion into Skiddaw Slates and the steeply inclined cleavages of the slates. It is suggested that the intrusion is the site of a former volcanic centre from which lavas and tuffs would have been erupted.
11. 235 153, Rigghead Quarries. These old slate workings (bedded tuffs) can be visited by those wishing to complete a circular tour ending at Grange.
12 and 13. 229 158 to 229 159, Newland Beck. The Borrowdale Volcanics, dipping 30° south-east, rest on vertical Skiddaw Slates trending north-south. At locality 13 the slates are sharply folded with the same north-south trend and although there can be no doubt that an unconformity exists here, there is equally no doubt that a single cleavage affects both the slates and the volcanics (Jeans, 1972).
14 and 15. 231 162 and 231 166, Eel Crags. Conglomerate similar to that of localities 7 and 8 occurs at the base of the crags (Moseley, 1975; Wadge, 1978).
16. 229 170. The Skiddaw Slates on the crag above the old mine form the main feature of interest but their interpretation is for experts only. Exceedingly complex and small-scale polyphase folding is exhibited.

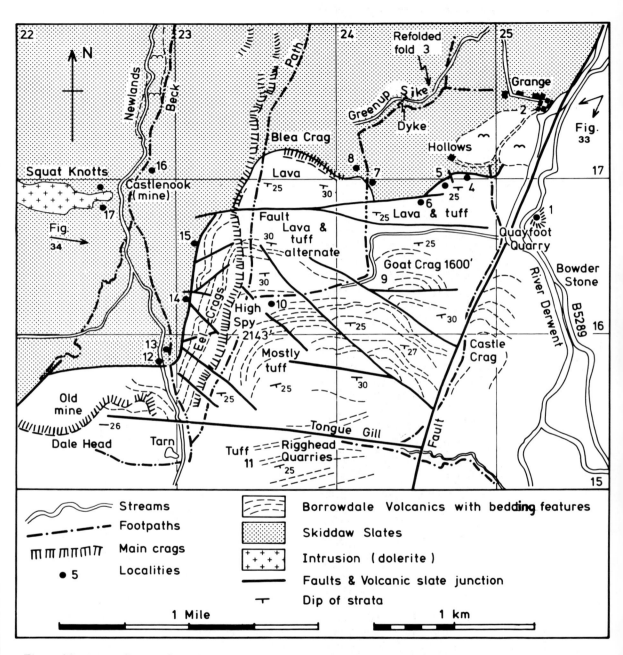

Figure 32 An outline geological map of Borrowdale, High Spy, Newlands area (excursion B). The viewpoint for figure 33 is indicated at NY 256 175 and that from figure 34 at NY 222 166. There are complex alternations of lava and tuff across this region which cannot be shown on a map of this scale, nor can be detailed structures of the Skiddaw Slates be shown

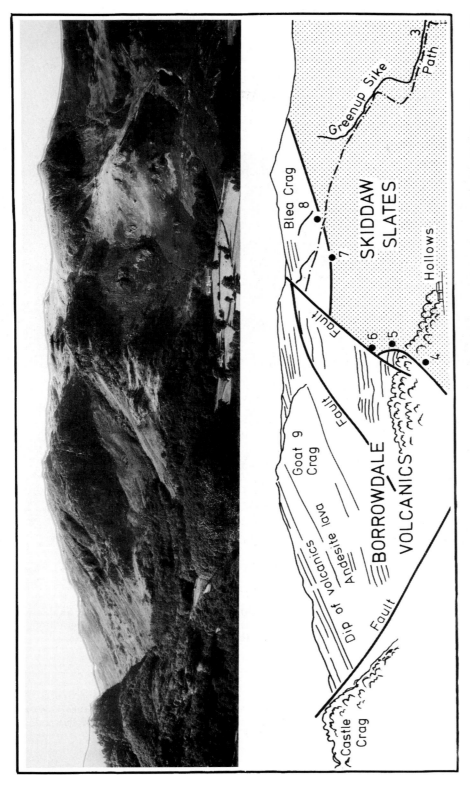

Figure 33 A view of the Hollows Farm — High Spy area from west of Grange (NY 256 175, see figure 32). The Skiddaw Slate — Borrowdale Volcanic junction is easily seen from this position and localities referred to in excursion B are marked

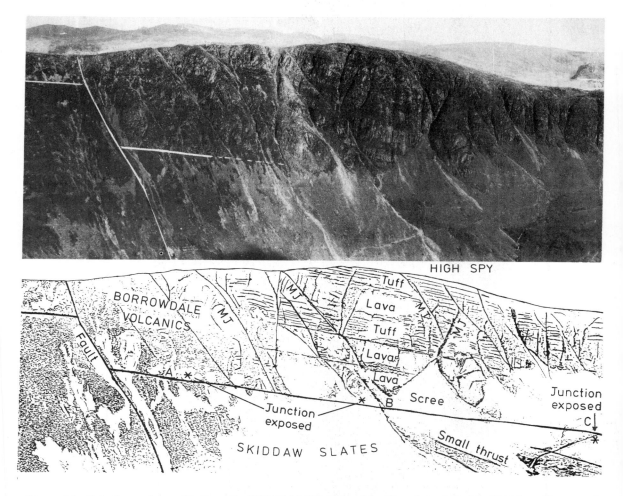

Figure 34 Eel Crags from Hindscarth (NY 218 168). The junction between the Skiddaw Slates and Borrowdale Volcanics can be seen with the volcanics rising to the top of the crag as alternations of lava (the massive parts of the crags) and tuffs (broken crags with more vegetation)

Figure 35   Dale Head and Hindscarth from High Spy (see figures 32 and 34). S — Skiddaw Slates with cleavage (C). V — Borrowdale Volcanics with bedding (B). VF is a faulted slice of volcanics within the slates and J is the approximate junction between slates and volcanics. M — old copper workings of Long Work vein. D — Squatt Knott basic intrusion

17. 225 169. Part of the Squat Knotts dolerite (to diorite) intrusion can be inspected just above the west bank of Newlands Beck.

A number of exposures of the slate–volcanic junction have been described. It is my opinion that this junction represents a distinct unconformity, the slates having been subjected to north–south folding before the volcanic rocks were formed; but I do not think that this represents a major orogeny, and it is probable that the cleavage of the slates was formed in post-volcanic (end-Silurian) times.

C. ST. JOHN'S, HIGH RIGG AND BRAMCRAG (1 : 25 000 maps NY22, NY32)

This is a small compact area; it is covered by dividing the excursion into two sections, one to the west and the other to the east of St. John's Vale. The western section is of comparatively low relief; although the highest point of High Rigg at 1000 feet (300 metres) rises no more than 300 feet (90 metres) above the starting point near St. John's Church, the topography is varied with open fellsides and a scattering of small steep crags. It is delightful walking country with magnificent views towards Skiddaw and Thirlmere. The geology is so varied and interesting that a whole day can easily be occupied by an excursion that involves only a very little walking or climbing. This excursion is therefore ideal for a hot summer day. The eastern section is dominated by the much steeper cliffs of Wanthwaite and Bramcrag, and contrasts scenically with High Rigg. The topographic scale is larger and the fellsides sweep away across the northern reaches of the Helvellyn Range.

The region is interesting because of the great variety of rocks and structures that can be seen in a small area. These include lavas and tuffs of the Borrowdale Volcanics, small outcrops of Skiddaw Slates and the St. John's (Threlkeld) Microgranite which has been intruded close to the junction between the slates and volcanics.

The Skiddaw Slates are not well exposed but it can be demonstrated that they (a) underlie the microgranite with a nearly horizontal contact (east of the Vale of St. John) and (b) overlie the microgranite (Low Rigg). Although it has not yet been possible to determine the detailed shape of the microgranite, reconstructions suggest that it is lens shaped and it has been interpreted as a laccolith. The volcanic rocks above consist of basaltic andesite and andesite lavas with thin tuff bands; such rocks are well exposed on both sides of the valley.

The western area can be explored from St. John's Church, where cars can be parked (NY 306 224). Just to the west of the church there is an open route to the felltop, which is only 300 feet above. The localities that can be conveniently visited are as listed below (see figure 36).

1. 308 223. A short walk across the fellside brings one to an excellent viewpoint for Wanthwaite and Bramcrags (figure 37Y). The bedding in the volcanics, the junction between the volcanics and the Skiddaw Slates and Threlkeld Microgranite outcrops can all be seen. This is the area east of St. John's Vale that will be referred to later.

2. 303 223. Retracing one's steps to the bottom of the crags, an interesting volcanic succession can be examined at leisure. The lowest exposures (in the lowest crag) are red breccias (fragmental deposits) which are believed to be flow-brecciated andesite or basaltic andesite lava. The red iron oxide colour suggests that subaerial weathering occurred shortly after eruption. Bedded tuff forms the bench above this lava and rests unconformably upon it, indicating a considerable pause in volcanic activity. Above this are more lavas and tuffs as indicated on figure 38. The top of the crag is the viewpoint for figure 36. The St. John's Microgranite and its junction with the Skiddaw Slates are readily seen from this point.

3. 303 220. 250 metres to the west of locality 2, along the crest of the crags and across a deep fault depression, there are several intrusive breccias that have been interpreted as explosion breccias (Moseley, 1977). One of these breccias has a pipe-like form, the others are dykes (figure 39);

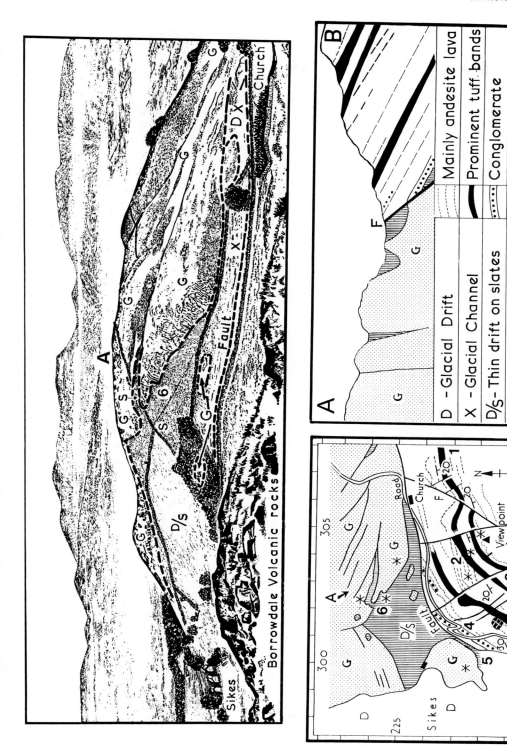

D – Glacial Drift	Mainly andesite lava
X – Glacial Channel	Prominent tuff bands
D/s – Thin drift on slates	Conglomerate
3* – Interesting Locality	G Microgranite
F – Fault ⤢ Dip	S Skiddaw Slate

Figure 36 The sketch is of the view from B (High Rigg) towards Low Rigg, with Skiddaw and Saddleback in the distance. Localities 1–6 mentioned in the text are indicated as is the line of the section A–B. Compare with figure 37X which is a view of High Rigg from near locality 6

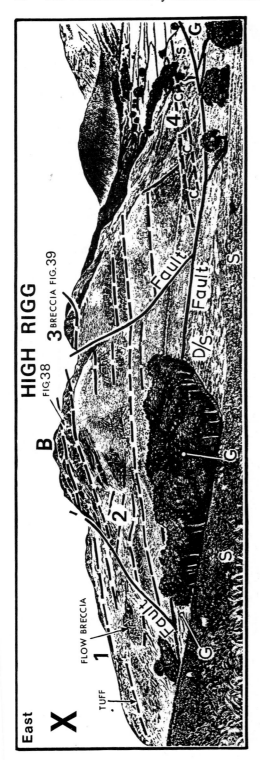

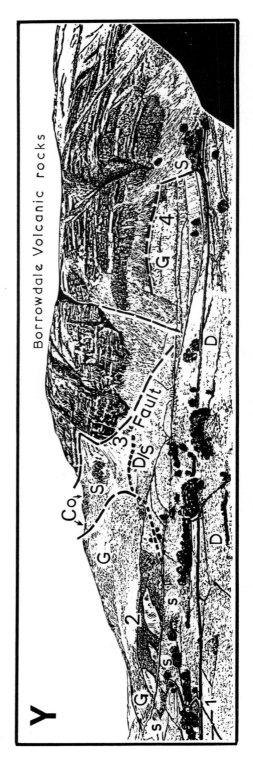

Figure 37 (X) A view of High Rigg from near locality 6 (figure 36). B is also the viewpoint for figure 36. Locality numbers mentioned in the text are shown. (Y) A view of Wanthwaite and Bramcrags from NY 308 223, with locality numbers shown. Gently dipping volcanics are faulted against a complex zone of Skiddaw Slates (Co), and there are two outcrops of microgranite (G); the Bramcrag intrusion (locality 4) is overlain by thin sediments whilst the Threlkeld Microgranite (locality 2 and figure 40) is intruded into the top of the Skiddaw Slates (Holland, 1981)

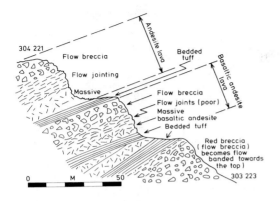

Figure 38 The succession of lava and tuff near to locality 2 on High Rigg (figure 36)

they penetrate a strongly flow-jointed andesite lava. Interpretation of these exposures requires careful and detailed study.

4. 301 223. A steep descent should now be made to the track running west from St. John's Church. In the zig-zag as the track drops down the hill there is a well-defined conglomerate that is believed to be near the base of the Borrowdale Volcanics. It consists of rounded clasts (pebbles and boulders), mostly of volcanic material but with occasional pieces of Skiddaw mudstone. This suggests that when the conglomerate was formed the volcanic eruptions were already underway and various early lavas had been eroded and water worn; also that the underlying Skiddaw Slates (mudstones) were exposed at the time and subject to active erosion.

5. 300 222. On the other side of the track there is a stile that gives access to good exposures of the St. John's Microgranite (part of the same complex as the Threlkeld Microgranite). There are also several small outcrops on the south side of the microgranite where contacts with Skiddaw Slate can be seen. The latter have been bleached and hardened by heat from the intrusion; they weather to the same colour as the microgranite so that their presence can be determined only by careful inspection of the outcrop.

6. 304 225 to 301 226. In order to examine other microgranite outcrops it is necessary to walk back along the road to the gate close to St. John's Church. It is then possible to follow the edge of the microgranite which forms such a feature here (figure 36). Here there are more contacts between microgranite and slate, and at 301 226 one has the view of High Rigg and of localities 1 to 4 referred to above (figure 37Y). The area of Low Rigg is close to the top of the microgranite intrusion as shown on the section on figure 37X.

The eastern side of St. John's Vale differs topographically from the High Rigg–Low Rigg area and although the geology is similar, with Skiddaw Slates, microgranite and volcanics present, the exposure of these rocks and their field relations differ considerably. In detail the geology of this region has not been completely explained, but the fundamentals are illustrated by figure 37Y. Skiddaw mudstones (now slates) were followed by lavas and tuffs of the Borrowdale Volcanics with the laccolithic sheet of the Threlkeld Microgranite intruded along the junction towards the end of the volcanic episode. During the Caledonian Orogeny the enormous stresses that crumpled much of the Lake District resulted in the resistant, strong (competent) volcanic rocks being thrust over the soft, weak (incompetent) Skiddaw Slates along a plane inclined 45° south (the fault at locality 3, figure 37Y).

1. 316 231. A good starting point for this excursion is at Wanthwaite where there is some parking space.

2. 320 230. A track leads to Hill Top Quarry (figure 37Y) where there are good exposures of the Threlkeld Microgranite. Unusual folded joints occur here; they are open recumbent folds, probably related to intrusive processes during injection of the microgranite rather than to the compressive forces of the Caledonian Orogeny.

3. 325 224, Buck Castle. To proceed from locality 2 to locality 3 it is best to walk south along the Bramcrag track, thus avoiding the direct route and an awkward wall which may be damaged if it is climbed.

The junction between volcanics and Skiddaw

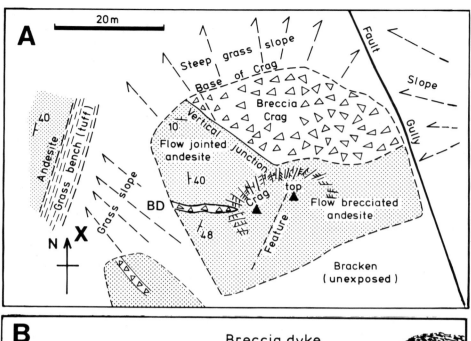

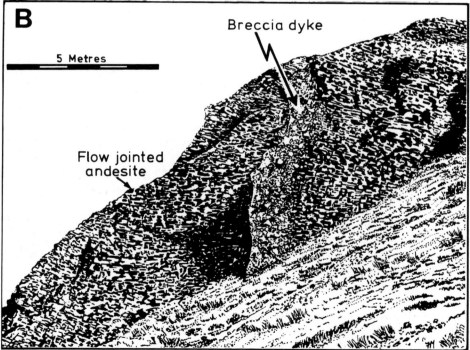

Figure 39 **(A)** A detailed map of the intrusion breccias of locality 3 (NY 303 220 and figures 36 and 37). The main breccia is composed of angular blocks of lava and is a pipe-like body intrusive into andesite lava. There are also two breccia dykes (BD) of similar composition. **(B)** A sketch of the breccia dyke BD from position X

Slates is exposed at the base of Wanthwaite Crags and takes the form of a mineralised fault plane dipping 45° south.

4. 320 220. An easy walk downhill takes one to Bramcrag Quarry where there is much of interest. The quarry is in an isolated offshoot of the Threlkeld Microgranite, which is overlain by about 10 metres of sandstone and siltstone (at the top of the quarry and inaccessible to all but a climbing party) and then by the Borrowdale Volcanic tuffs and lavas of Bramcrag. (Some interesting fallen blocks of porphyritic lava and coarse tuff can be inspected on the quarry floor.) At the south end of the quarry the microgranite is faulted against Skiddaw Slates. An age date of 445 million years has been determined for the microgranite (Wadge *et al.*, 1974), which sug-

gests that it was intruded towards the end of the Borrowdale Volcanic episode (see figures 40 and 41).

D. ULLSWATER TO KIRKSTONE PASS

A traverse from the shores of Ullswater to Kirkstone Pass and Ambleside covers the entire volcanic sequence from basalt and basaltic andesite flows, which form the earliest volcanic events in this region, to airfall tuffs and acid lavas near the top of the volcanic pile. This traverse has been divided into sections, each of which makes an interesting excursion: Gowbarrow, Barton Fell, Hallin Fell, Place Fell, and Kirkstone Pass to St. Sunday Crag

Figure 40 The succession of lava (escarpments mostly) and tuff (benches) forming Bramcrag, Vale of St. John. The Bramcrag microgranite quarry is on the left of the photograph

(see figure 42). The highest part of the succession exposed near Ambleside on Wansfell Pike is omitted because these rocks are more difficult to interpret.

Da. Gowbarrow (1 : 25 000 maps NY32, NY42)

Most of the localities referred to here were visited during a Geologists' Association excursion (Capewell, 1954; Moseley, 1964; Moseley *et al.*, 1972) and a similar route to the one followed on that occasion is shown on figure 43, starting at Birk Crag and ending at Airy Force. All these rocks belong to the Ullswater Group, the lowest division of the volcanic sequence in this region. Individuals and small groups operating from a single car may find it more convenient to subdivide the itinerary as indicated below so that starting and finishing points are the same place.

1 and 2.   434 215 and 432 216. These localities are on and adjacent to the minor road to Watermillock.

Locality 1 has excellent views of Barton Fell to the south-east (figure 42) and Birk Crag to the north-west. In both cases lava flows result in strong escarpments. Nearby at locality 2 there are several interesting old quarries. The first of these shows a junction between tuffs of the Borrowdale Volcanics and 'pencil slates' of the underlying Skiddaw Slates. The junction at this point is a fault but interpretation is made difficult by the weathered state of the outcrop. The other quarries are in coarse tuffs cut by a single vertical cleavage, a result of the Caledonian Orogeny (see chapter 4). Bedding in these tuffs is inclined about 20° north-east, and can be seen in a few places where there are compositional changes from coarser-grained to finer-grained tuffs.

3. 430 217. An ascent of Birk Crag crosses a number of andesite lava flows which have been tilted by earth movements to 25° to the north-east, similar to the tuffs of locality 2. This dip is shown by the flow jointing which is parallel to the tops of the flows. There is an excellent

Figure 41   A view from Threlkeld towards the south showing the Borrowdale Volcanics of White Pike (V), the Skiddaw Slates (S) and the Threlkeld Microgranite (G). The quarry is on the right of the photograph

Ullswater panorama from here which takes in Barton, Hallin and Place Fells (figure 42).

4. 426 223. Just beyond Hagg Wood there are excellent exposures of steeply dipping bedded tuffs forming the same part of the sequence as the tuffs of locality 2.

5. 427 229. A well-defined path crosses the upper part of Priest's Crag where there are good exposures of strongly flow-jointed dark basaltic andesites with dips of 50° north-east. Excellent views are obtained from here of the thrust-faulted junction with the Skiddaw Slates which is marked by the base of the crags extending south to Birk Crag and Knots.

6. 429 240. North of Priest's Crag there are outcrops of the much younger Devonian Conglomerates of Little Mell Fell which can be examined in a small quarry near the Folly. The conglomerate is made of rounded sandstone boulders, some of which are more than a foot in diameter. It is here cut by an intrusion of amygdaloidal basalt.

It is at this point that the excursion can be split into either a return to locality 1 or a walk across Great Meldrum to locality 7.

7. 402 224. The crag south-west of Ulcat Row displays characteristic flow brecciation in andesite lava with the fragments recessed by weathering. There is also some flow banding and flow folding (figure 13).

8. 395 220, Norman Crag. The andesite lava here displays excellent flow brecciation, resembling that of locality 7.

9. 403 219. At about 1400 OD (Ordnance Datum) on Gowbarrow Fell, andesitic ignimbrite with eutaxitic texture is exposed on a small knoll. This rock is easily mistaken for flow-banded andesite lava.

10. 408 218. The summit of Gowbarrow Fell (Airy Crag) is formed from one of several thin andesite flows which result in a well-defined 'trap topography'; that is, the lavas form escarpments and the more easily eroded tops and bottoms of the lavas form ledges.

11. From Airy Crag there is a pleasant walk across Gowbarrow Fell with good views of the south-eastern side of Ullswater from 410 208.

12. The walk ends in the Airy Force region (figure 45, 400 206). The stream here is partly controlled by joints in andesite lava and the waterfall may have receded from the volcanic-Skiddaw Slate junction. Downstream the junction can be seen at 400 205 (a high angle fault) and further downstream there are andesitic intrusions into the Skiddaw Slates (402 201). There is a large car park at 400 200.

Db. Barton Fell, Ullswater (figure 46) (1 : 25 000 map NY42)

It is easy to appreciate the volcanic sequence of Barton Fell from distant views such as those illustrated by figures 42 and 48. It will be noticed that a prominent fault brings the volcanic rocks against the Skiddaw Slates — this is easily seen from the topography — with the soft slates forming gently sloping ground and a number of escarpments (lava flows) completely truncated by the fault. The effect of this fault is to cut out the earliest of the volcanic rocks, which are therefore not seen here at outcrop although comparison with adjacent regions leaves little doubt that the lower part of the Barton Fell sequence is an approximate correlative of those of Priest's Crag and Hallin Fell (Da and Dc).

From a distance the rocks appear to be nearly horizontal, but closer inspection reveals that this is not so and that the whole sequence has been tilted and now dips to the south-east at about 45° (figure 47). This region in fact forms one limb of the Place Fell syncline, an important structure formed during the Caledonian earth movements (see excursion Dd).

Localities 1-8 represent a convenient excursion route across the Barton Fell sequence but there are alternatives; for example, a straight traverse along the line of the section of figure 47 is equally instructive.

1. 458 211, Auterstone Crag. The prominent gully shown on figure 48 provides an interesting section as follows. At the bottom of the gulley there is garnetiferous andesitic breccia (coarse

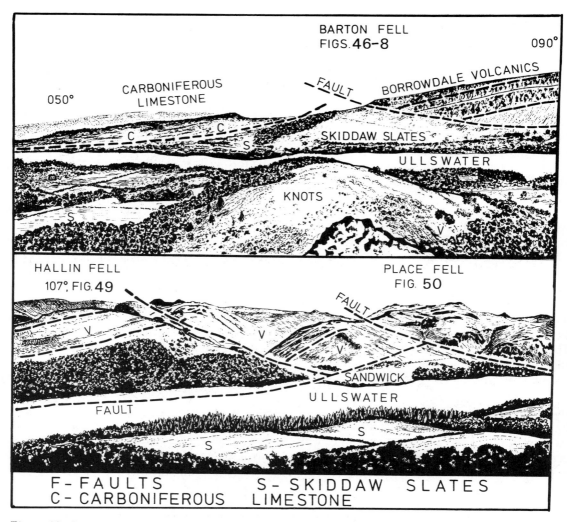

Figure 42   Panoramic view of Ullswater from Birk Crag (NY 430 216). The excursion areas of Barton

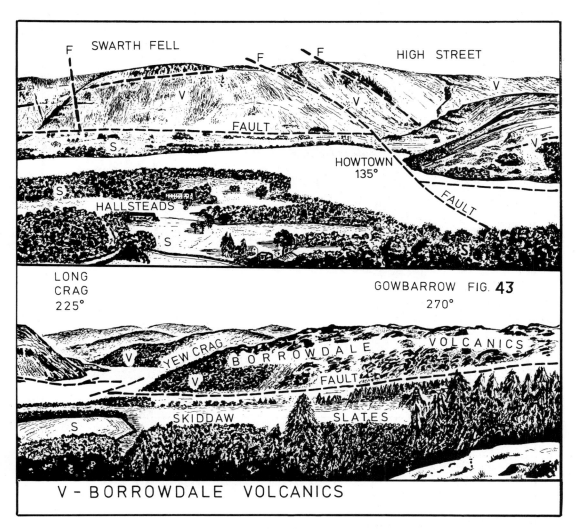

Fell (Db), Hallin Fell (Dc), Place Fell (Dd), and part of Gowbarrow (Da) can be seen

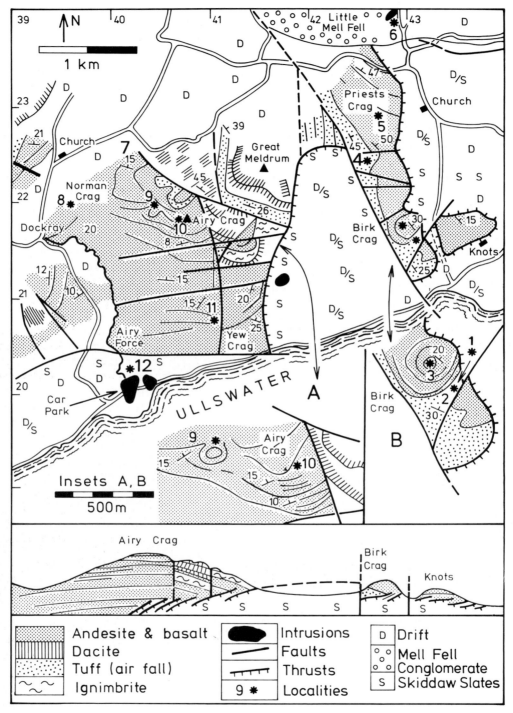

Figure 43 Geological map of the Gowbarrow area (excursion Da and figure 44). The section from Airy Crag (Gowbarrow Fell) to Knots shows particularly the low angle (thrust) fault contact between the Skiddaw Slates and the volcanics

tuff). This is overlain by garnetiferous andesite lava and then by more tuff. The presence of the mineral garnet (small red crystals) is interesting and is one indication that the original magmas were deep seated; that is, they were probably derived from the mantle beneath the continental crust (figure 5). The tuff beds above the lava tend to form ledges between the lava escarpments.

2. 459 212. To follow this route it is necessary to return to the bottom of the gully (locality 1) and follow the base of the crag. Amygdaloidal andesite outcrops at the top of a grassy slope. This is lava with numerous small gas holes now filled with black chlorite and other minerals (figure 11).

3. Immediately below the andesite at the base of the crag there is fine-grained bedded tuff with some coarse and medium tuffs interbedded.

4. 459 213. The lower part of the crag is formed of breccia with large angular fragments. Below it there is bedded tuff in fine, medium and coarse bands.

5. 461 214. The base of the crag is medium-grained tuff; above it there is coarse breccia (flow breccia), and higher still comes amygdaloidal and perlitic andesite lava. Perlitic texture consists of small concentric cracks, which are believed to have developed during the cooling of a former volcanic glass. A hand lens is necessary to see this texture.

6. 463 214. Bedded tuff followed by massive porphyritic andesite lava can be traced along the fellside to the south-east. The tuff is easily eroded and forms a bench that is largely scree-covered whereas the lava forms a prominent escarpment rising towards the top of the fell.

7. 466 214. The escarpment here consists of two andesite lava flows. The lower one has developed flow joints and the upper one exhibits perlitic and amygdaloidal texture (see above). There are also bedded tuffs hereabouts which form benches between escarpments.

8. 469 216, White Knott. A variety of rocks are exposed on this crag including tuff and massive, flow-brecciated and perlitic andesite.

Dc. Hallin Fell (figure 49) (1:25 000 maps NY41, NY42)

The area between Sandwick and Howtown on the south-east shore of Ullswater makes an excellent excursion for those not wishing to be too strenuous. It is scenically attractive with numerous rocky outcrops but with none of the steep crags and peaks so exhausting for the over eighties, and there are excellent views across the lake from Hallin Fell. Geologically this area provides one of the best sections across the lower part of the Borrowdale Volcanics; there is a thick sequence of basalt and basaltic andesite lava followed by andesite lava and tuff (Moseley, 1960). All these rocks have been tilted to angles of about 50° by the Caledonian earth movements. For convenience the localities are described from a starting point at Sandwick (NY 423 198), where Skiddaw Slates are exposed. The excursion ends on Birkie Knott, south-east of St. Peter's Church (438 188).

1. 425 196. Skiddaw Slates and Volcanics exposed in Sandwick Beck are almost certainly faulted against each other. The lowest exposure of volcanics consists of coarse-grained and medium-grained tuffs with mudstone fragments (derived from the Skiddaw Slates). A short distance upstream (426 195) a basalt flow is cut by quartz veins.

2. From Sandwick follow the footpath north-east along the lake shore. At 428 201 there are coarse-grained greenish tuffs containing fragments of basaltic andesite.

3. 431 203. Also on the lake shore there are basalt (or basaltic andesite) flows with close-spaced 'flow' jointing.

4. 434 204. Kailpot Crag is an historic exposure described by Marr (1916a); a pothole just above lake level is ascribed to the action of glacial melt-waters. Coarse-grained to medium-grained tuffs are exposed here.

5. 436 204. Further along the lake path at Geordies Crag there is flow-banded basalt with occasional small garnets. The upper part of the

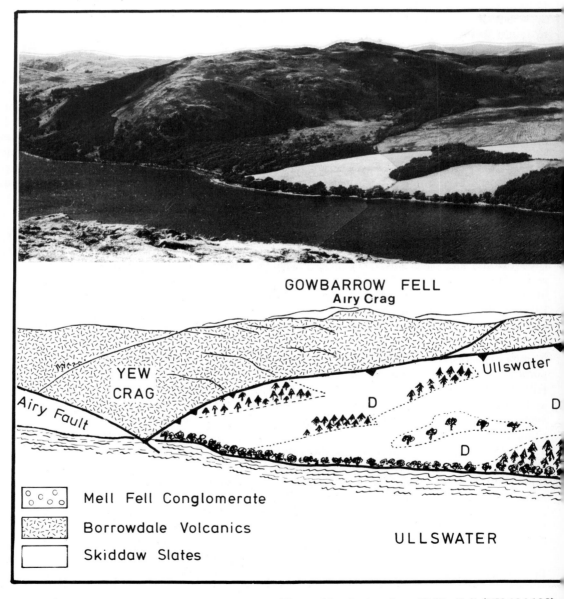

GOWBARROW FELL
Aıry Crag

YEW
CRAG

Airy Fault

Ullswater

D

D

D

Mell Fell Conglomerate

Borrowdale Volcanics

Skiddaw Slates

ULLSWATER

Figure 44   A view from Hallin Fell (NY 434 198)

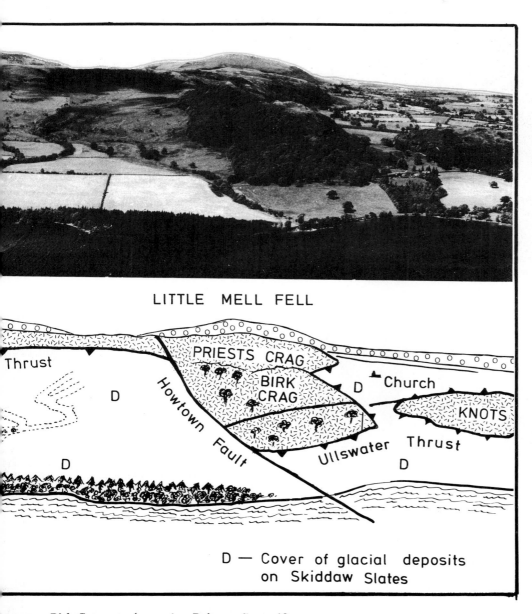

LITTLE MELL FELL

Thrust

Howtown Fault

PRIESTS CRAG

BIRK CRAG

D

D — Church

KNOTS

Ullswater Thrust

D

D

D

D — Cover of glacial deposits
on Skiddaw Slates

arrow — Birk Crag area (excursion Da); see figure 43

Figure 45   Airy Force, Gowbarrow (NY 400 206). The waterfall is partly controlled by joints in andesite lava

flow is reddened (haematite) suggesting subaerial extrusion. It was at this locality that Professor W. B. R. King, 30 years ago, admonished me for unnecessary hammering, a lesson we should all learn.

6. To see the sequence to the best advantage it is now necessary to leave the path and head uphill (strongly bracken-covered in summer). At 433 202, immediately above the boundary wall of Hallinhag Wood, there is a sequence consisting of coarse-grained, medium-grained and bedded tuffs followed by basalt lava, the base of which is partly flow brecciated and partly flow banded.

7. 430 200. This locality is at Swine Slacks and requires a diversion from the direct route. The sequence consists of unbedded tuff with a variety of sedimentary structures, followed by flow-banded basalt with platy flow joints. The top of the flow is brecciated.

8. 431 198. At the top of the crag 250 metres west of Hallin Fell summit there are basalt lava flows with platy 'flow' jointing.

9. 434 200. An escarpment 250 metres NNE of Hallin Fell summit reveals reddened (subaerial?) basalt lava, and just above is a columnar jointed flow.

10. 433 198. At Hallin Fell summit there is basaltic flow breccia followed by normal basalt lava.

11. 436 200. 250 metres north-west of Hallin Fell summit there is a lenticular outcrop of tuff. It may represent a small volcanic vent.

12. A well-defined path leads from Hallin Fell east to Martindale Hause. At 435 195 it crosses basalt lava with well-developed flow joints and small phenocrysts of feldspar and altered pyroxene.

13. Reverting to the lakeside footpath around Hallin Fell, a distinctive feature at 439 200 runs up the fellside. It consists of medium-to-coarse tuff with fragments of variable lithology, including pumice, and with some reddening. The band is underlain and overlain by basalt lava with feldspar phenocrysts and altered pyroxene.

14. 439 192. Two to three hundred metres east of St. Peter's Church there is flow-banded andesite with perlitic texture. It is underlain by andesitic and felsitic tuff.

15. 438 189. Crags 150 metres south-east of Lanty Tarn exhibit bedded andesitic tuff with 'bird's eye' structures (figure 16).

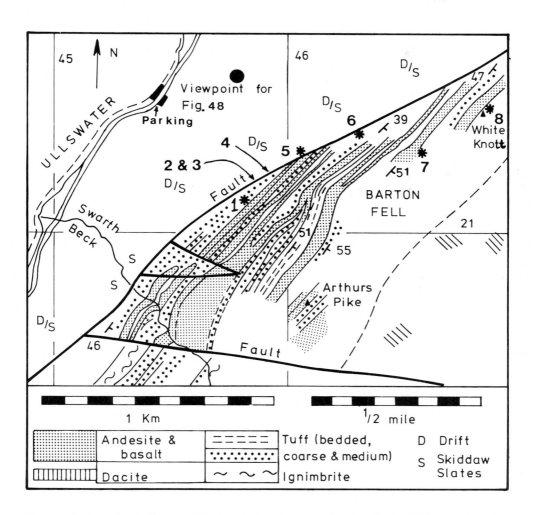

Figure 46   A geological map of Barton Fell, Ullswater, showing the localities mentioned in excursion Db

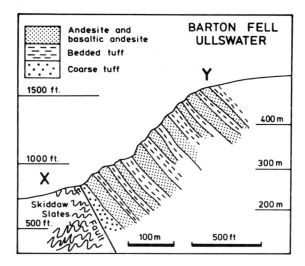

Figure 47 Section across Barton Fell. The lava flows form the escarpments and are separated from each other by more easily eroded tuffs, which form benches and are often scree-covered. The volcanic sequence has been tilted to an angle of 40-50° (the north-west limb of the Place Fell Syncline) and is faulted against Skiddaw Slates

Dd. Place Fell (1:25 000 maps NY 31, NY 41)

The volcanic sequence of Place Fell (Moseley, 1960) has some similarities to those of Gowbarrow (Da), Hallin Fell (Dc) and Barton Fell (Db), since like those areas it is the lowest part of the volcanic sequence that is exposed. However, the unpredictability of volcanic eruptions, with successive lava flows and pyroclastic deposits invading quite different areas, is so great that the differences in the sequences are more pronounced than the similarities, although the general pattern of basaltic andesites followed by normal andesites and acid volcanics is preserved.

An excellent excursion to cover this ground starts and ends at Patterdale. It is scenically beautiful, geologically fascinating and follows the route indicated on figure 50 along established footpaths to Silvery Point and Birk Fell, leaving the path for scree and rough ground between Long Crag and Place Fell summit and eventually arriving back at Patterdale on another footpath. It is an energetic excursion, affording excellent views of the surrounding country, that requires a full day. It should not be a race, otherwise there will be appreciation of neither the scenery nor the geology. For those who wish for a less strenuous route a pleasant alternative is to follow the path from Long Crag via Hallin Fell to Howtown (Dc), where a return to Glenridding can be made on the lake 'steamer'.

From Patterdale there is a track leading to Side Farm. The exposures of volcanic rock start near here and will be described under locality numbers (figure 50). It should be understood that these are localities picked out as possibly the most interesting; there are many others in intermediate positions.

1. 397 165. The wood immediately north-west of Side Farm has moderately cleaved, bedded andesitic tuff with one horizon of 'bird's eye' tuff (figure 16).
2. 396 166. A small crag at the south-west extremity of Ullswater exhibits moderately cleaved volcaniclastic andesitic tuff with sedimentary structures that indicate deposition in water.
3. 398 168. Just above the upper footpath at between 800 and 900 feet OD there are outcrops of strongly porphyritic basaltic andesite with phenocrysts of pink feldspar (albite on analysis) and irregular green aggregates (epidote and penninite).
4. 397 169. Between the footpaths, but extending up the fellside, there are well-bedded and strongly cleaved andesitic volcaniclastic tuffs that have been worked for slate (several small quarries). This band can be traced across the fell to High Dod.
5. 393 175. At Purse Point on the lake shore there are most interesting outcrops of flow-banded, nodular and flow-brecciated 'dacite' (see also figures 13, 14 and 59). The latter has bedded tuff-like deposits filling the cracks between lava blocks.
6. 396 180. It is best to take the higher path which crosses the col between Silvery Crag and Birk Fell (Bleaberry Knott). To the west of the col a dark basaltic andesite lava with feldspar and pyroxene phenocrysts is followed by a thin sequence of coarse tuff, bedded tuff, ignimbrite

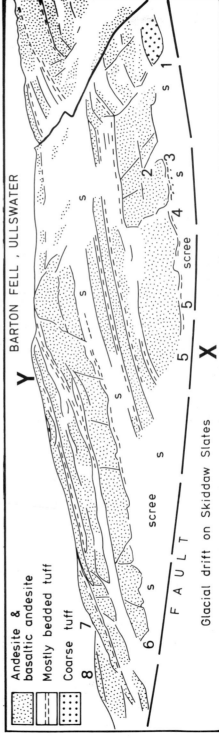

Figure 48 A view of Barton Fell from the west. The resistant lavas forming escarpments are easily seen as is the approximate position of the fault that cuts obliquely across several lava flows. Locality numbers 1–8 on figure 46 and the position of section X–Y on figure 47 are marked

The following labels appear on the geological map (Figure 48):

BARTON FELL, ULLSWATER

FAULT

scree

Glacial drift on Skiddaw Slates

Legend:
- Andesite & basaltic andesite
- Mostly bedded tuff
- Coarse tuff

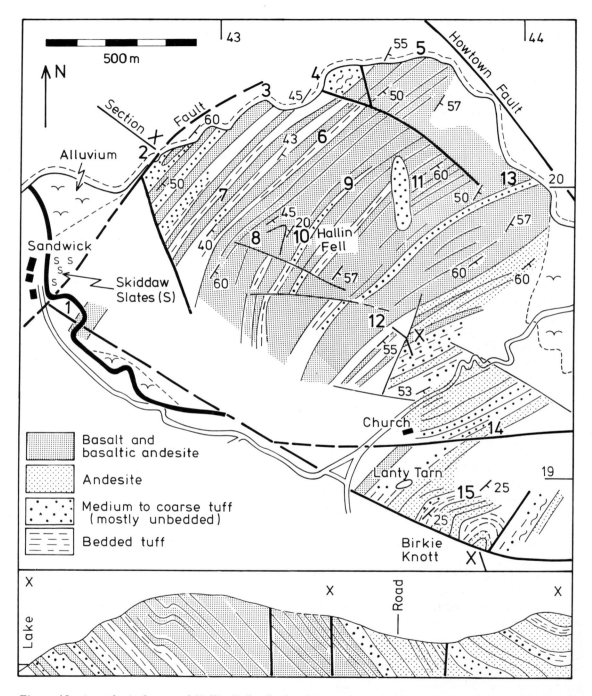

Figure 49   **A geological map of Hallin Fell. The locality numbers 1–15 referred to in excursion Dc are marked**

(with eutaxitic texture), silicified bedded tuff (hornstone) and flow-banded 'dacite'.

7. 395 184, Silvery Point. Outcrops from the lake shore to the crag top, 100 feet higher, reveal several strongly flow-brecciated andesite flows separated from each other by thin layers of bedded tuff. Weathered surfaces of the flow breccia show typical recessed fragments (figure 13). A little higher 396 183 there is an andesite lava with some flow banding.

8. 410 194. To follow the itinerary listed here it is necessary to leave the footpath at Long Crag and climb the scree alongside the crag. Between 700 and 750 feet OD there are porphyritic basaltic andesite lavas with platy flow joints and numerous garnets.

9. Continuing across Long Crag there are excellent views of the Gowbarrow region (see excursion Da) and near Scalehow Force (415 191) there are strongly flow-brecciated andesites.

10. Further upstream at the base of a small waterfall (412 186) an altered olivine basalt dyke can be seen intruded into pale 'dacite'.

11. An interesting westerly diversion (406 184) leads to an interesting exposure of 'pumice' between 'dacite' flows.

12. 409 182. South of Scalehow Beck there is a lenticular bed of eutaxitic ignimbrite interbedded with andesitic tuff.

13. A thick, easily recognised dark basaltic andesite unit is crossed. It extends from High Dod to the lake shore near Side Farm (see locality 3).

14. 406 170. At the south-west end of Hart Crag immediately north of Place Fell summit there is a section as follows. Coarse tuff with andesitic and felsitic fragments outcrops at the base of the crag (this may have resulted from a mudflow (lahar)); above this there is basaltic andesite lava with some flow brecciation. Immediately south of Place Fell summit there are felsitic tuffs and flow-brecciated 'dacite' lavas.

15. Between localities 14 and 15 the axis of the Place Fell Syncline is crossed. This structure can be traced across the whole of the Lake District and in the west is known as the Scafell Syncline. At 15 (411 169), on lower ground above Hawk Crag, there are pronounced dip slopes in bedded tuff which here is on the south-east limb of the syncline. South-west from locality 15 a footpath leads back to Patterdale.

De. Kirkstone Pass to Fairfield and St. Sunday Crag (1 : 25 000 maps NY 30, NY 31)

This excursion involves a long walk; ideally transport should be left at Kirkstone Pass and arrangements made for it to be rejoined at Patterdale. If this is not possible it will be necessary to walk back along the road to Kirkstone Pass (figure 51). Alternatively other excursion routes can be made up by following the locality numbers in different orders; for example, Patterdale – Grizedale – St. Sunday Crag – Deepdale, or Deepdale – Fairfield – Dove Crag – Little Hart Crag – Brothers Water. It is an excellent area in which to study a thick sequence of tuff, mostly airfall but partly volcaniclastic, together with occasional ignimbrite flows and dacite, andesite and basalt lava. The walk across the higher fells affords excellent views, with the geology clearly evident in distant crags. The starting point, at 1500 feet, gives a good height advantage but there are nevertheless a few stiff climbs.

1. 402 075 to 398 080. About 500 metres south of the Kirkstone Inn on the Windermere road there is an excellent view towards the west of the slopes of Kilnshaw Chimney and Pets Quarry. The rock in Pets Quarry is volcaniclastic tuff (deposited in water and reworked (figure 52)) and is important as an ornamental stone of considerable export value. The depositional layers (or bedding) are easily seen and are gently inclined. They can be traced across the fellside to Kirkstone Pass (figure 53). At Kirkstone Pass itself there is a well-formed deposit of hummocky moraine left behind by the glaciers of the last ice age.

Immediately to the west of the pass on the lower slopes of Kilnshaw Chimney, bedded tuff can be examined that is similar to that worked in Pets Quarry.

2. 395 088. Between Kirkstone Pass and the top of Kilnshaw Chimney, 1000 feet higher, the sequence consists mostly of coarse tuff although there is a small amount of dacite and andesite lava.

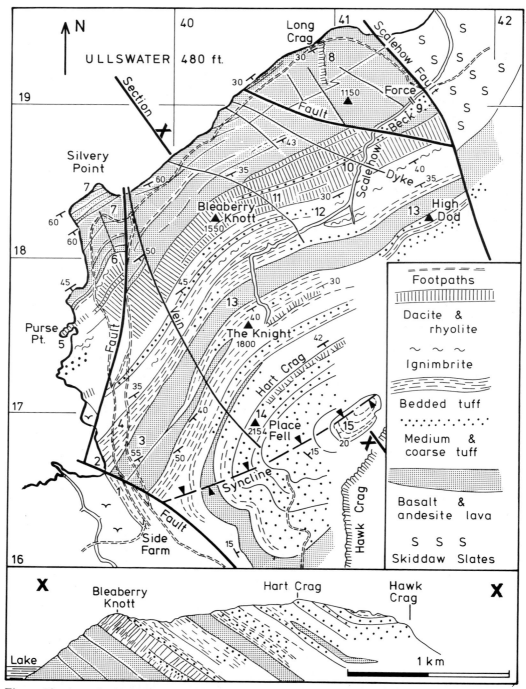

Figure 50   A geological map of Place Fell. The locality numbers 1–15 referred to in excursion Dd are marked. The Place Fell Syncline can be traced across the whole of the volcanic outcrop as far as Scafell (see figure 26)

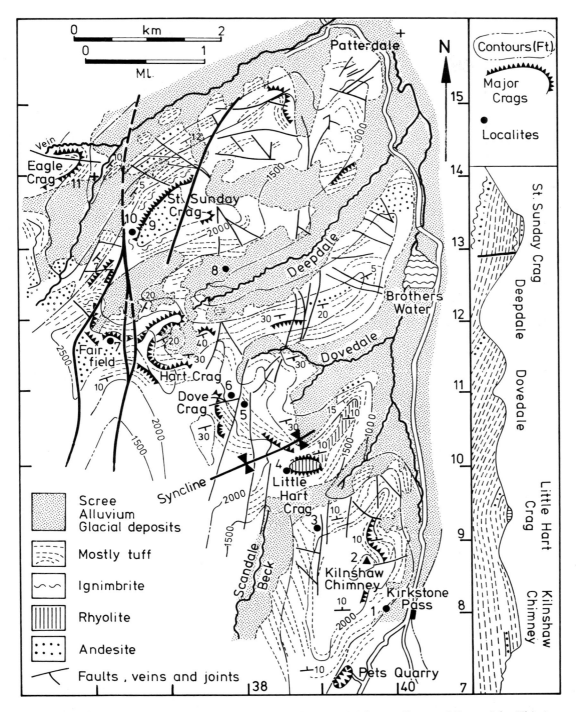

Figure 51 An outline geological map of the area between Kirkstone Pass and Patterdale. This is a region dominated by airfall tuffs, with subsidiary lava flows and ignimbrite (excursion De)

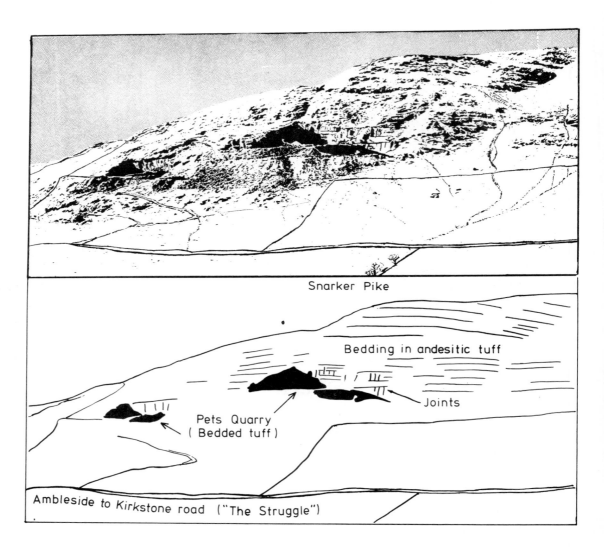

Snarker Pike

Bedding in andesitic tuff

Joints

Pets Quarry
(Bedded tuff)

Ambleside to Kirkstone road ("The Struggle")

Figure 52 A view to the west from the Kirkstone — Windermere road just south of Kirkstone Pass (NY 401 076). The volcanic succession here consists essentially of water-lain bedded tuffs, which are quarried for their ornamental value. They are followed, above the quarry, by coarse tuffs that have not yet been investigated

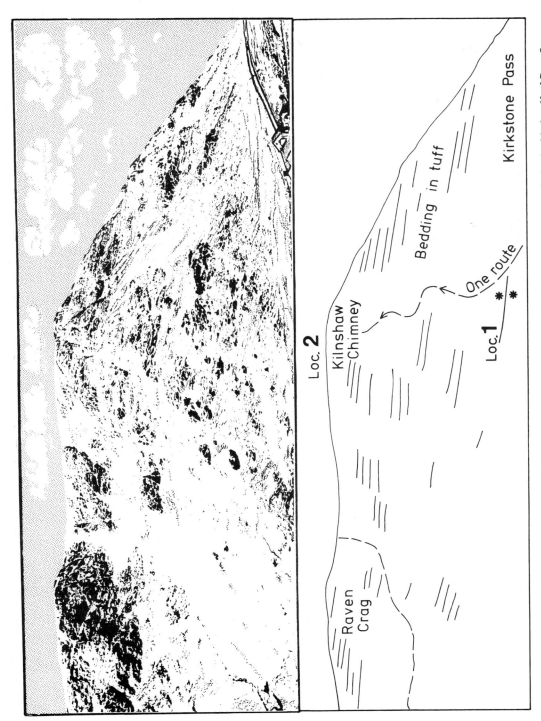

Figure 53 The view from above the Kirkstone Inn (excursion De). The succession (Loc. 1) starts with the bedded tuff of Pets Quarry (figure 52); this is followed in turn by coarse tuff, dacite lava (about halfway up the crags, bedded tuff (there is one beautifully cross-bedded unit) and coarse tuff, to Loc. 2)

Kilnshaw Chimney is rather flat on top with thin deposits of peat and little rock exposure.

3. 389 092. To the north-west the ground slopes gently to the col between Kilnshaw Chimney and Little Hart Crag. It is mostly made up of dip slopes in bedded tuff; that is, the dip of the strata and the slope of the ground are parallel to each other.

4. 385 100. Little Hart Crag is a complex outcrop situated close to the axis of the Haweswater Syncline, where the dip changes from a NNW to a SSE direction. It is composed of dacite lava with a variety of flow banding and flow folding (figure 15). The latter, formed during the emplacement of the lava, has resulted in the banding being locally inclined at high angles.

5. 379 109. A gentle walk from Little Hart Crag takes one to the undulating ground below Dove Crag. There are excellent views of the Dovedale region to the north-east, where the rocks are predominantly tuff but with occasional andesite flows. The bedding is easily seen running along the fellsides. To the north-west there is the impressive precipice of Dove Crag, well known to rock climbers. This is also nearly all tuff with the bedding inclined 30° to a southerly direction. There are a number of prominent gullies which are small faults and master joints (figure 54). At locality 5 there are exposures of coarse tuff and agglomerate with some fragments several feet in diameter, suggesting the nearby presence of an Ordovician volcanic vent.

6. 377 110. Except for suitably experienced rock climbers with aspirations to some of the vertical routes, the Dove Crag sequence should be examined by traversing round the northern end, eventually to end up on the high ground between Dove Crag and Hart Crag. There is an easier walk from Little Hart Crag to the west of Dove Crag but it is not recommended since there is little to see of geological interest.

7. 361 118. A gentle walk across broad flat tops takes one from Hart Crag to Fairfield. The rocks are mostly bedded tuffs forming north-facing cliffs but a flow of basalt (or basaltic andesite — it has not yet been analysed) forms the summit area of Fairfield. The sequence (figure 55) can be inspected by descending one of the gullies into the upper part of Deepdale.

8. 376 127. The overlook from Fairfield into Deepdale is worth contemplating. Deepdale is a good example of a glaciated valley, providing an excellent example of hummocky moraine in the region of locality 8. There is a good footpath down Deepdale to the road for those wishing to cut the excursion short.

9. 365 134. The route now continues along the ridge via Cofa Pike to St. Sunday Crag. The flat top is composed of a gently dipping basalt lava, with excellent views of Fairfield to the south (figure 54), and Eagle Crag and the Helvellyn Range to the north-west.

10. 364 133. A descent of the precipice of St. Sunday Crag should on no account be contemplated except by those experienced in rock climbing. It is easy enough, however, to traverse round the western end of the crag and thus examine the sequence illustrated by figure 56. There are varieties of tuff exposed but the most notable feature is the ignimbrite sheet, which has been described by Millward (1980); it has the characteristic eutaxitic texture (dark streaks in a paler matrix) that has been described elsewhere (figure 20).

11. 357 144. The easiest descent is to the north into Grizedale. A good view will be obtained of the Eagle Crag Vein which penetrates gently inclined tuffs (figure 57). A footpath then leads back to Patterdale.

E. KIRKSTONE PASS TO KIDSTY PIKE (1 : 25 000 map NY 41)

The immediate area of Kidsty Pike has recently been described by D. Millward (Millward, 1980), but the surrounding area is not well known and no recent publications are available. Kidsty Pike, 2500 feet high, is an offshoot of High Street and can be reached in several ways. As far as I am aware, one of the geologically most interesting routes to follow is from Kirkstone Pass although the return trip involves a moderately long walk. A shorter but still interesting route is from the

Figure 54 Dove Crag from the east with the suggested route for excursion De shown. It often requires a practised eye to pick out geological structures, and that is the case here. There are, however, a number of places where bedding and faults can be seen

Figure 55   Greenhow End, Fairfiled (NY 365 120) from St. Sunday Crag (NY 370 135). Hart Crag and Dove Crag are beyond. The volcanic sequence here consists almost entirely of bedded tuffs with some coarse tuffs. In this view the bedding is inclined gently from right to left

Figure 56 St. Sunday Crag (NY 365 135) seen from the north-west. The succession starting at the top is: a, basalt lava; b, bedded and coarse tuff; c, ignimbrite; d–e, coarse tuff and bedded tuff; f, bedded tuff; g, dacite lava. The ignimbrite unit, which is best inspected in the section at h, is particularly interesting and has been described by Millward (1980). The crag is truncated by a major fault (figures 26 and 51) which crosses the whole of the volcanic outcrop from Coniston to Ullswater

head of Haweswater via Blea Tarn (a perfect example of an ice-eroded hollow or corrie formed at the height of the ice age) to Riggindale Crag (the viewpoint for figure 58). From here Kidsty is reached via High Street.

The most interesting part of Kidsty Pike is the summit area. The south face, over 200 metres high, consists entirely of pyroclastic rocks with an interesting ignimbrite sequence near the top (figure 58). This sequence starts with a pale-grey rubbly rock which only an expert would be able to identify as one variety of ignimbrite. It is followed by a unit about 6 metres thick which is nodular at the base and top, and massive with some ignimbritic streaks (fiamme) in the middle (figures 19 and 20). This has been interpreted as a single cooling unit of an ignimbrite flow (see figure 59). The nodules, subspherical and silica-rich, are believed to have formed largely as a result of volatile (vapour) release as the flow cooled (Millward, 1980). Above the upper nodular layer there is

another ignimbrite which contains garnet and has numerous dark lenticular streaks (fiamme), generally referred to as a eutaxitic texture. David Millward interprets this as one cooling unit made up of several flow units. The suggestion is that a number of incandescent flows followed one another in rapid succession, and then cooled together as one unit.

The alternative walking route from Kirkstone Pass to Kidsty Pike is about 10 km in length but it does have the advantage of starting at 1500 feet, only 1000 feet lower than Kidsty Pike. There is however the little matter of Threshthwaite Cove into which one has to drop nearly 1000 feet before climbing out again on the other side. The geology along this route is interesting and not unlike that on the western side of Kirkstone Pass (excursion De). It is dominated by gently dipping bedded tuff and coarse tuff, with structures clearly seen on distant fellsides (compare with figures 52 and 53).

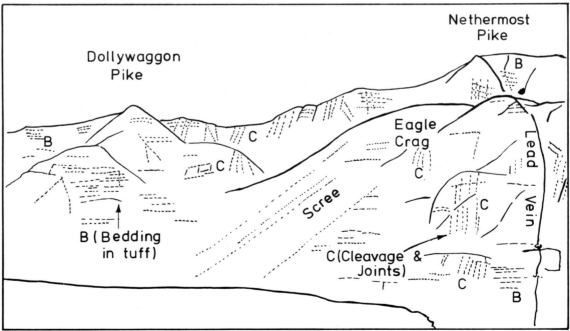

Figure 57 A westerly view from Grizedale showing the Eagle Crag vein which was formerly worked for lead. The volcanic rocks are mostly bedded tuffs (figure 17, etc.)

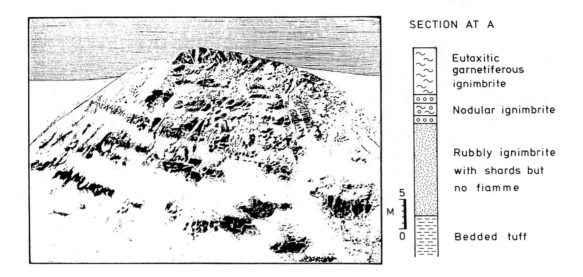

SECTION AT A

Eutaxitic garnetiferous ignimbrite

Nodular ignimbrite

Rubbly ignimbrite with shards but no fiamme

Bedded tuff

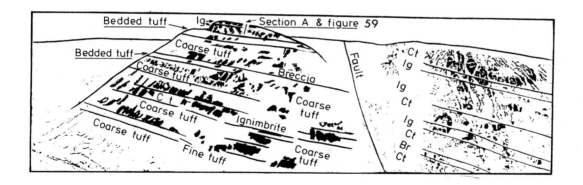

Figure 58   Kidsty Pike from Riggindale Crag (NY 445 113). The most interesting section is at A in the summit area and has been described by Millward (1980). See the text for greater detail

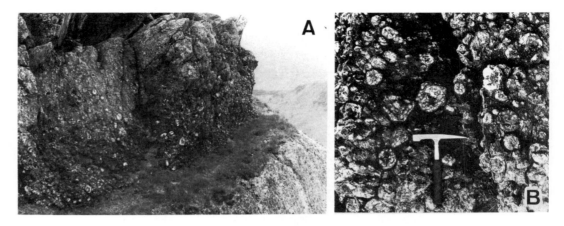

Figure 59   Kidsty Pike. (A) The lower nodular bed of the Kidsty Pike ignimbrite with the rubbly ignimbrite beneath, see figure 58, section at A. (B) The nodules (subspherical) are believed to have formed *in situ* during the cooling of the rock as a late-stage vapour phase. If further details are required consult Millward (1980)

F.   LANGDALE (figure 60)

The region surrounding the Langdale Pikes is one of the most popular in the Lake District. Climbers swarm over the impressive cliffs of Gimmer, White and Raven Crags and Pavey Ark, and walkers abound on all the fell paths from the gentle amble to Dungeon Ghyll waterfall to the more strenuous fell walks across the tops and to Scafell and Borrowdale. This region provides excellent examples of volcanic geology that can be seen and understood easily from distant views of the crags, as indicated in the sketches and in more detail from close up inspection of the rock outcrops. However, readers are asked not to deface the outcrop by using geological hammers; good specimens can be collected from adjacent scree slopes. Most of the Langdale rocks were formed by particularly violent eruptive processes and *nuées ardentes* (ignimbrites) were common (see chapters 1 to 3). A local volcanic sequence more than 800 metres thick was built up, with the oldest rocks outcropping near the valley bottom. None of the rocks have been seriously deformed by folding and although there are numerous faults, these are mostly of small magnitude such as those shown on figures 61 and 62.

The Airy's Bridge ignimbrites and dacites (rhyolites)
This group of rocks forms many of the fellsides from Langdale to Great Gable and southern Borrowdale. It has been estimated to be more than 1000 metres thick by Oliver (1961), who derived this name from Airy's Bridge near Styhead. In upper Langdale these rocks outcrop from the valley floor nearly to the top of the higher crags of Pike O'Stickle, Gimmer Crag and Thorn Crag, although on the lower slopes there are concealing deposits of moraine and scree resulting from the action of glaciers of the ice age and from subsequent erosion. Further west there are extensive outcrops on the Band and Crinkle Crags. The rock sequence is by no means simple however, and in detail is much more complex than that illustrated diagrammatically by figure 60.

At the bottom there are dacite and rhyolite lavas; the distinction between these two high-silica rocks depends on their detailed chemistry and this cannot be determined in the field (chapter 3). They are followed towards the top of Raven and White Crags by ignimbrite flows of similar composition, which must have been erupted with great violence rather like the disastrous *nuée ardente* of Mont Pelée, Martinique in 1906 and more recently

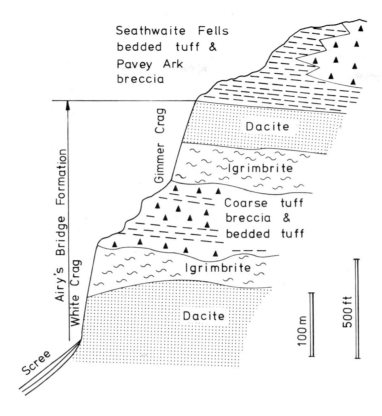

Figure 60 The geological succession in the Langdale district

the great 1980 blast from Mount St. Helens in the Cascades of north-west America. The Airy's Bridge ignimbrites, like those of present-day volcanoes, indicate that the local volcanic centre stood above the sea at this time.

The ignimbrites are followed abruptly, still within the Airy's Bridge Group, by coarse tuffs and bedded tuffs, the latter with beautiful sedimentary structures which indicate deposition in water with current and wave action subsequently redistributing the ash (volcaniclastic). Structures of this kind are referred to in the descriptions of figures 17, 18 and 66. The abrupt change from ignimbrite to tuff shows that there were considerable changes in the volcano at this time; perhaps there was strong local subsidence and erosion with a moderately long time interval between eruptions of the ignimbrites and tuffs, not unexpected when comparison is made with modern volcanoes.

Higher up the fellsides a return to dacite lava and ignimbrite can be seen in the steep rock of Pike O'Stickle, Gimmer Crag and Thorn Crag, and extending round the flank of Harrison Stickle to just below Stickle Tarn (figure 61). The volcano was thus elevated into a subaerial environment once again. Small faults, which formed much later during the main Caledonian mountain building (chapter 4), have been eroded into prominent gullies (figures 62 and 63).

Seathwaite Fells Tuffs
The highest outcrops on the Langdale Fells consist of a sequence of bedded tuffs very similar to those just described. These rocks form the upper part of Pike O'Stickle, Gimmer Crag, Thorn Crag and Harrison Stickle and extend further afield to Bowfell, Scafell and into southern Borrowdale. They are andesitic tuffs that were laid down in

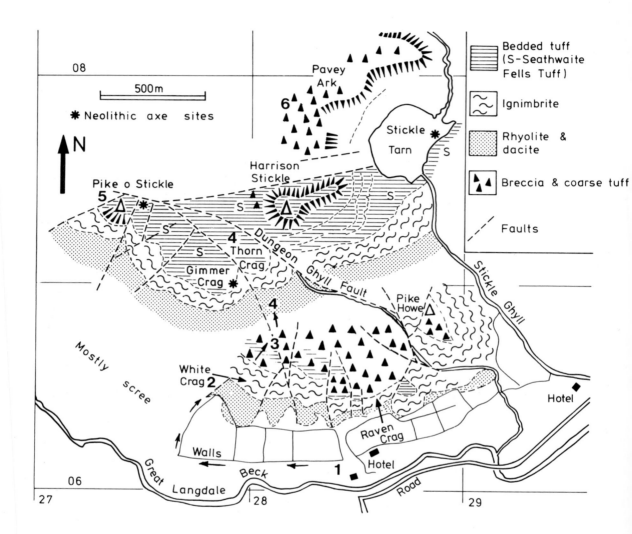

Figure 61 Outline geological map of the Langdale Pikes with a suggested excursion route indicated by numbers 1–6 and arrows

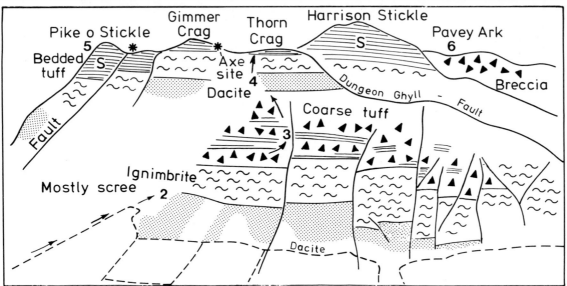

Figure 62 The Langdale Pikes seen from the south (NY 285 054), with a suggested excursion route indicated (see figure 61)

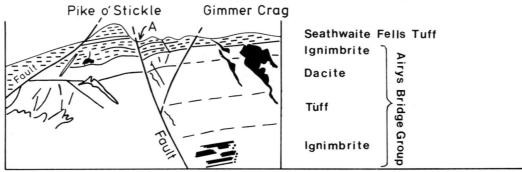

Pike o'Stickle   Gimmer Crag

Seathwaite Fells Tuff

Ignimbrite
Dacite
Tüff
Ignimbrite

Airys Bridge Group

Fault

Fault

A

Figure 63 Pike O'Stickle seen from The Band (NY 260 060). Note that the faults following the main gullies are quite small, although on figure 62 they appear large. This is because of the perspective of the view on that figure and shows that one must be careful not to interpret too much from a single viewpoint. Location A is a Neolithic axe factory site

Figure 64 **The lower part of the Seathwaite Fells Tuffs on Gimmer Crag. Harrison Stickle is beyond, also in Seathwaite Fells Tuffs**

water and redistributed by wave and current action. The lowest few metres are of particular interest because the outcrop was worked by Stone Age man for axe manufacture (figure 64). There are many axe factory sites between Stickle Tarn and Scafell, all situated in this same band of rock. But in case the reader imagines he can organise collecting sorties, it must be emphasised that the most common artifacts are flakes chipped off the axes, since the completed axes were 'exported' to other parts of Britain (conceivably Britain's first export). Flakes are likely to be distinguished from accidentally broken rock only by an expert. Reverting to the rocks themselves, they are commonly composed of closely spaced alternations of fine-grained and very fine-grained silicified ash (bedded tuff) (figure 64). The sudden and widespread change from the ignimbrite below was a major event in the history of the volcano, probably indicating a major subsidence beneath

the sea, followed, perhaps after a considerable lapse of time, by renewed eruption of andesitic ash from another volcanic centre.

Pavey Ark Breccia

The one other easily identified volcanic rock in the Langdale area is the breccia centred on Pavey Ark (figure 65). It is of local occurrence and is best seen on the upper parts of Pavey Ark where it is made up of large angular volcanic fragments thrown together in the chaotic mix. It partly cuts across and lenses into the Seathwaite Fells Tuff outcrop, and represents a violent explosive event from a nearby volcanic centre. At the northern end of Pavey Ark (at the base of the cliff) dark-grey andesite lavas can be seen.

Several interesting excursion routes can be followed in the area covered by the Langdale Pikes, Bowfell and Crinkle Crags, as follows.

Figure 65 (A) Volcanic breccia (agglomerate) near the top of Pavey Ark (NY 283 078), excursion Fa, locality 6. This outcrop is composed of large irregular shaped fragments embedded in a finer matrix. These were formed by volcanic explosion and must have been deposited close to one of the vents of the former volcano. (B) Near-vertical flow banding and jointing in dacite lava on Crinkle Crags, excursion Fd, locality 7

Fa. Langdale Pikes (1 : 25 000 map NY20)

A good route to follow is shown on the map (figure 61) and on the Langdale views (figures 62, 63 and 67). Start at the Old Dungeon Ghyll Hotel (NY 286 061) and take the footpath towards Rossett Ghyll.

1. It is worth examining loose boulders along this path; excellent examples of flow-banded dacite lava and ignimbrite, which have fallen from the crags above, will be seen.
2. Leaving the footpath at 276 062, follow Grave Gill up the fellside (along the wall seen on the left of figure 62). This route crosses outcrops of flow-banded and brecciated acid lavas (dacites) of White Crag and the ignimbrites above them (Moseley and Millward, 1982).
3. Bedded and coarse tuffs still higher up the fell-side are well exposed on small crags. Some of these rocks have structures of considerable interest and it is well worthwhile wandering slowly over this ground. A beeline for the top of the fell will result in much being missed. Descriptions of hand specimens of these and similar rocks are given in chapter 3.
4. From the top of White Crag the route continues to the base of Thorn Crag (280 070), where it meets the path from Dungeon Ghyll. There is a similar sequence to that seen on White Crag with dacite lavas at the bottom of Thorn Crag followed by well-developed ignimbrites and, at the top, beautifully bedded tuffs belonging to the Seathwaite Fells Tuffs (figure 66). Climbers ascending routes on Gimmer Crag will cross the same sequence (figure 62).
5. The Seathwaite Fells Tuffs are well exposed from Pike O'Stickle to Harrison Stickle, but note the interbedded breccia bands towards the top of Harrison Stickle.
6. Continuing along the path from Harrison Stickle to Pavey Ark the Pavey Ark Breccia is soon encountered (283 076). It consists of a chaotic assemblage of angular blocks, the largest several metres in diameter, and is underlain, towards the bottom of Pavey Ark, by gently dipping bedded tuffs and by andesitic lava. This breccia was probably deposited close to one of the vents of the Borrowdale Volcano.

The easiest way back to the road is via Stickle Tarn and Stickle Ghyll but a more interesting route is to skirt round the contours to Pike Howe (figure 67).

Fb. Pike O'Blisco (1 : 25 000 map NY20)

There are no recent geological maps of the area covered by this excursion (and also excursions Fc and Fd). Hence the observations recorded represent reconnaissance traverses only.

The variety of rocks on Pike O'Blisco is similar to that on the Langdale Pikes but the field relations are quite different, so that this excursion is by no means a repetition of that of the Pikes. An excellent place to start is at the col (cattle grid) on the road from Langdale to Blea Tarn (see figures 68 and 69). The col corresponds to a major shatter zone (fault) extending from Rosset Ghyll to the gully behind Blea Tarn House. The displacement of this fault is not known but it is probably small.

1. 289 051. Cross the swampy patch south-west of the road. The first outcrop south of the old wall (about 40 metres from the road) is pink dacite lava. Above it there is bedded tuff and coarse tuff.
2. 289 050. Cross the wall to the top of the ridge; here bedded tuff is followed by pink dacite, the latter outcropping along the ridge to the south-west.
3. 287 048. Follow the ridge until a north-west–south-east depression is reached. This depression probably coincides with a small fault. On the north-eastern side of the depression there is a porphyritic dacite with flow structure and on the extreme south-west side there is dark bedded tuff. Sixty metres to the north-west another outcrop of flow-jointed dacite will be seen and from here a small path leads out of the depression in a direction of 225° (this path starts at the col, locality 1). It is best to follow this path; although it eventually peters out, carry on in the same direction of 225° up a steep grassy gully that

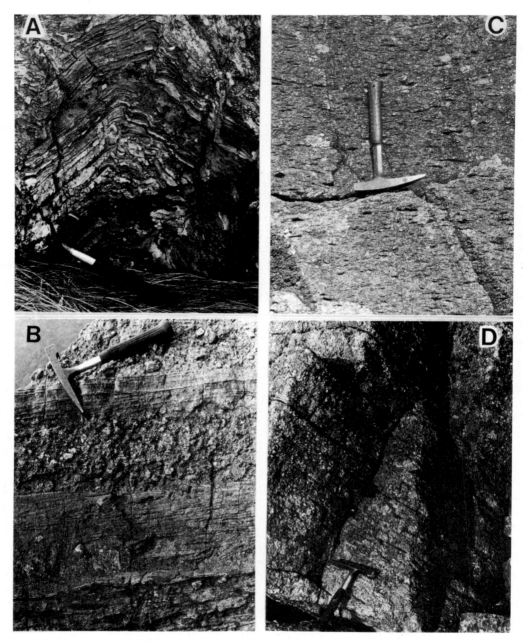

Figure 66   (A) Seathwaite Fells Tuffs on the top of Thorn Crag (NY 280 070 and figures 61 and 62), excursion Fa, locality 4. This outcrop displays a great variety of water-lain sedimentary structures. In this case the folding was caused by slumping of wet sediment, and not by the much later tectonic (mountain-building) process (compare with figure 21). (B) Bedded andesitic tuffs alternating between coarse grain and medium grain, Pike O'Blisco, excursion Fb, locality 7. There is a great variety of depositional sedimentary structure to be seen in these outcrops. (C) Ignimbrite with a strong eutaxitic texture (lenticular streaks), The Band, excursion Fd, locality 1. (D) Flow-banded dacite lava (mostly flow joints insofar as the rock breaks along closely parallel planes), Crinkle Crags, excursion Fd, locality 5

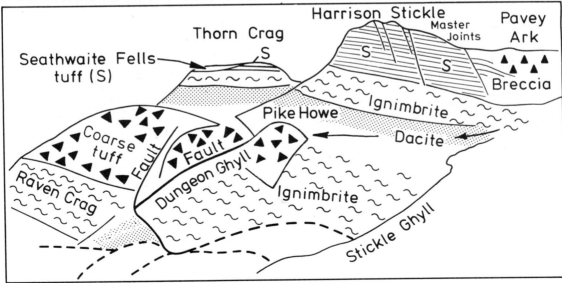

Figure 67 The mountainside rising above the New Dungeon Ghyll Hotel. The return journey from the suggested excursion descends the upper part of Stickle Ghyll and skirts the mountainside to Pike Howe

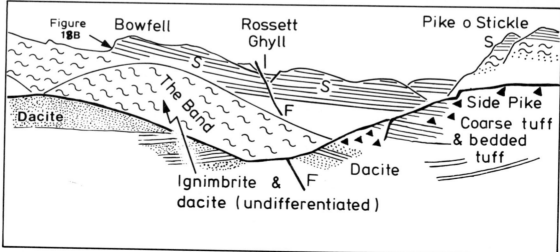

Figure 68 A view from near Blea Tarn (NY 297 046) towards the Langdale Pikes and Bowfell; S, Seathwaite Fells tuff, F, fault

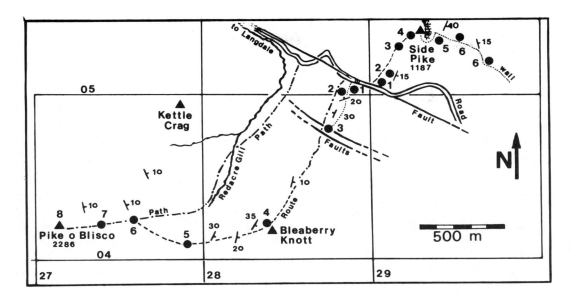

Figure 69 A map showing the excursion routes Fb and Fc

rises about 500 feet to the skyline of Bleaberry Knott (284 042). There are exposures along this route but the more interesting geology is beyond the top of the gully.

4 and 5. 284 042 to 279 041. Beyond Bleaberry Knott the ground is gently undulating, with almost continuous exposures of dacite and bedded tuff that have average dips of 20° south-east and alternate as follows:

F   Flow-banded dacite (stratigraphically the highest unit)
E   Bedded tuff with layers of coarse tuff
D   Dacite flow breccia with flow banding
C   Flow-banded dacite
B   Bedded tuff
A   Dacite

6. 276 043. Move across gently sloping ground towards the base of the final crags of Pike O'Blisco where the principal footpath will be reached. Here there are beautifully exposed bedding surfaces in bedded tuff, large slabs that are easily seen from a distance. This tuff dips about 10° in an easterly direction and is level-bedded with a variety of sedimentary structures.

7. 274 043. The path rises more steeply across small escarpments. Here there is coarse tuff and breccia (large angular fragments) and alternations of bedded tuff with a wide range of sedimentary structures (figure 66B).

8. 271 042. The final escarpment rising to the top of the Pike consists of coarse dacite breccia, probably flow-brecciated lava. There are large fragments recessed by weathering in an extremely fine-grained and partly flow-banded matrix.

Fc. Side Pike (1: 25 000 map NY 20)

This is a short excursion with the same starting point as excursion Fb (figure 69). Again lack of recent geological maps of the area make the observations represent reconnaissance traverses only.

1. 290 051. The lowest crag 50 metres from the road has a section as follows:

C   Pink dacite
B   Bedded tuff (3 metres)
A   Grey to pink flow-jointed dacite

2. 291 051. The second crag 100 metres from location 1 consists of coarse tuff followed by bedded tuff, forming a prominent scar heading in the direction of Blea Tarn.

3. 291 053. Higher up the fell there are alternations of bedded and coarse tuff, the latter with large fragments towards the top of the crag.

4. 293 054. The crag levels off, followed by a small drop before the final rise to Side Pike. At the crag base there is bedded tuff but the main part is andesitic flow breccia with the characteristic weathered surfaces of recessed fragments (figure 13).

5. 294 053. The eastern face of Side Pike is a vertical precipice, which can be by-passed by following a small path to the south. This leads to a col beyond which is bedded tuff and coarse tuff.

6. 295 053 and 297 052. The path follows a wall and crosses a well-developed sequence of bedded and coarse tuff, the former with a wide variety of sedimentary structures and the latter with fragments approaching a metre in diameter in one band.

Fd. Crinkle Crags (1 : 25 000 map NY 20)

There are two obvious starting points for an excursion to traverse Crinkle Crags. One is at Wrynose Pass (NY 277 028), which is at nearly 1300 feet so that the climb is moderate; the other is at the Old Dungeon Ghyll Hotel, Langdale (NY 285 060), which is only 300 feet above sea level and hence the excursion involves a stiffer climb, but compensation is provided by the very interesting geological features encountered on the way. As with excursions Fb and Fc, the observations represent reconnaissance traverses only, owing to lack of recent geological maps.

Langdale, The Band, Crinkle and the lower slopes of Bowfell

1. The Band. Follow the footpath from Stool End (276 057). Detailed information will not be given but there are excellent ignimbrite exposures in places (figure 66C).

2. Three Tarns, 248 060. There is a variety of rocks exposed in this area. Agglomerate with large angular fragments outcrops near the col (figure 15) as does dacite lava. A few hundred feet up the path towards Bowfell (249 062) there are excellent exposures of volcaniclastic bedded tuff with a wide variety of sedimentary structures (the Seathwaite Fells Tuff, figure 18).

3. 249 056. Follow the footpath south across the crags, crossing ignimbrite with some dacite. *En route* there is an excellent view to the north, clearly showing the bedded tuff outcrops that form the Churns of Bowfell (246 062).

4. 250 052. There are outcrops of dacite lava and ignimbrite but in the absence of any recent survey the field relations are not yet clear.

5. 249 049, Main summit. This area is formed of flow-banded dacite, with dips varying from 30° north to 20° east (figure 66D).

6. 249 047. There is a steep drop to a col where the rock is dacite lava.

7. 250 045. Near a minor summit there are some most interesting sections in flow-banded dacite lava (figure 65B). There appears to be a sharp east–west anticlinal structure with the flow banding vertical in places. At present my opinion is that this represents flow folding, but the area is in need of detailed investigation. Ignimbrite is exposed a little further south.

8. 250 044. Dacite forms the final crag where the path drops down to unexposed ground.

The route from Wrynose Pass
Immediately to the north-east of the pass there is a succession of andesite and bedded tuff. Subsequently follow the main footpath past Red Tarn, this area being predominantly covered by hummocky moraine of glacial origin.

From NY 265 039 there is an excellent view of the western crags of Pike O'Blisco. The rocks in these crags are similar to those described in excursion Fb, and are nearly horizontal. This route then reaches location 8 of the route just described.

G. CONISTON

This classical area includes the upper part of the volcanics and the overlying Ordovician and Silurian sedimentary rocks and thus marks the extinction of the Borrowdale Volcano. There were many other important geological events about this time, of which perhaps the most significant was the big gap in time between deposition of the volcanic rocks and the Coniston Limestone that rests upon them. This is known as an unconformity and represents many millions of years for which there is no geological record in the Lake District. Shortly after the Borrowdale Volcano became extinct, a period of earth movements uplifted and folded the rocks of this region. There followed substantial erosion during which the volcanic landscape was reduced to an undulating plain and would no longer have been recognised as a volcano, just as the Lake District cannot be recognised as a volcano today. After a lapse of many millions of years the Coniston Limestone was deposited in a tropical sea and now rests on a variety of volcanic formations, but it must be remembered that because of the earlier erosion these formations do not represent the final eruption of the volcano.

The most important of the early descriptions of this region were those of J. E. Marr during the last century, culminating in his book on the geology of the Lake District in 1916 (Marr, 1916a, 1916b). His work on the Coniston Limestone has recently been updated by K. J. McNamara (McNamara, 1979) whilst the volcanic rocks have been described by G. H. Mitchell (Mitchell, 1940), another important name in Lake District geology; also see figure 70.

Considering the volcanic rocks first, it is possible to examine a pile about 4 km thick between Coniston and the Duddon Valley, because the whole sequence was tilted to a high angle during the Caledonian earth movements. A short walk, for example from Coniston to Levers Water, enables one to cross several thousand metres of upended volcanic rocks (figure 71) including ignimbrites, rhyolites, andesites, volcaniclastic and airfall tuffs and mudflows.

GEOLOGICAL SUCCESSION NEAR CONISTON

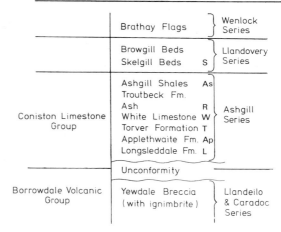

Figure 70 The succession near Coniston shown here is that proposed recently by K. J. McNamara (McNamara, 1979). Many of the formations are thin and not exposed over the whole area. It is a question of piecing together numbers of small exposures to make up the whole sequence. This takes time and requires an advanced geological mapping technique. The equivalent names used by Marr (1916b), where different from McNamara's names, are as follows: Troutbeck Formation (Phacops mucronatus Beds), Torver Formation (Phillipsinella Beds), Longsleddale Formation (lower part of the Stile End Beds; upper part not present in this area)

The overlying Coniston Limestone Group belies its name since only a pervert would immediately think of it as limestone. It is mostly lime-rich (calcareous) mudstone with some sandstone, shale and thin limestone bands, and has a shelly fauna of trilobites and brachiopods. These beds are followed abruptly by black deep-water Silurian shales, the Skelgill Beds, which contain abundant graptolites (figure 72). This rock unit is soft and easily eroded, and forms a conspicuous depression across the bottom of the fell (figures 73 and 74). Higher in the sequence there are the grey Browgill Beds (shales and mudstones) and then the grey striped Brathay Flags. Still higher Silurian formations will not be encountered on the Coniston excursions described here.

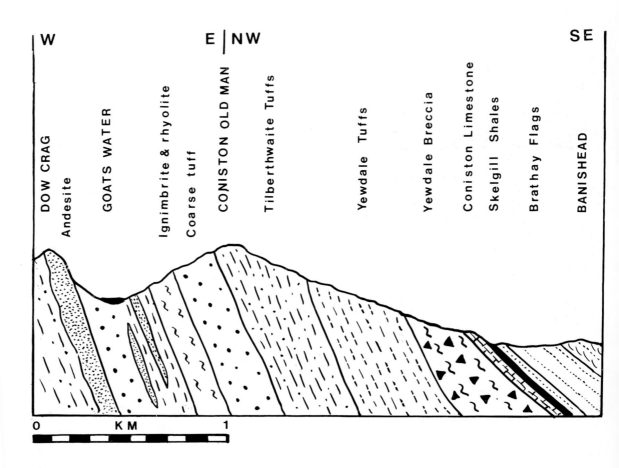

Figure 71   Section from Dow Crag to Banishead. The rocks were tilted to a high angle during the Caledonian earth movements (excursion Ga)

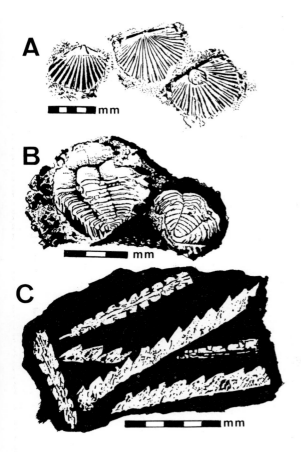

Figure 72 Fossils from the upper Ordovician (A, brachiopods; B, trilobites) and lower Silurian (C, graptolites)

Three excursions are listed of which the first (Ga) refers to both the volcanic and sedimentary rocks and the other two (Gb and Gc) entirely to volcanic rocks.

Ga. South-west Coniston (Timley Knott to Ashgill Quarry) (1 : 25 000 map SD 29)

Details of the geology for this excursion are shown on figure 73A, B with the most instructive localities indicated. The geology can be studied in either an elementary or an advanced way and it is a suitable area for all comers, providing sights are kept at the right level. I have, therefore, first considered those aspects that can be appreciated by those with little geological knowledge, and then dealt with items which require more experience.

Even for those with no previous knowledge of geology the following features should be understandable.

(a) Starting at Boo Tarn (282 986) where cars can be parked, a north-easterly view towards Timley Knott should leave no doubt about the approximate junction between the Borrowdale Volcanics and the Coniston Limestone (figure 75). The volcanics form the rough craggy ground rising towards Coniston Old Man.

(b) Boo Tarn occupies a linear depression which is followed to the north-east by Braidy Beck and to the south-west by the Walna Scar footpath. This is the outcrop of the soft, easily eroded Skelgill Shales.

(c) There is a low feature (a rise in the ground) immediately to the south-east of the 'Skelgill' depression. It is formed by the Browgill mudstones, which are slightly more resistant to erosion.

(d) There is an excellent outcrop of Brathay Flags in Banishead Quarry where both bedding and cleavage planes are clearly visible (figure 23A). This quarry is well worth a visit, especially when Torver Beck is in flood since a spectacular waterfall drops into it.

(e) Walk up the south-west bank of Torver Beck. Easily appreciated columnar joints in ignimbrite will be seen at 274 963 (see locality 9).

(f) A pleasant walk from here is to follow the path to Goats Water where numerous old slate quarries and the impressive Dow Crag can be seen (figure 76). This area provides excellent examples of glacial erosion and deposition.

I shall now describe the localities 1–18 shown on figure 73A, B. Some of the comments supplement those above and are appropriate for beginners; others are relatively advanced, suitable perhaps for university undergraduates.

1. 284 971, Timley Knott (figures 74 and 75). The details of the rock succession on Timley Knott are more complex than a first impression

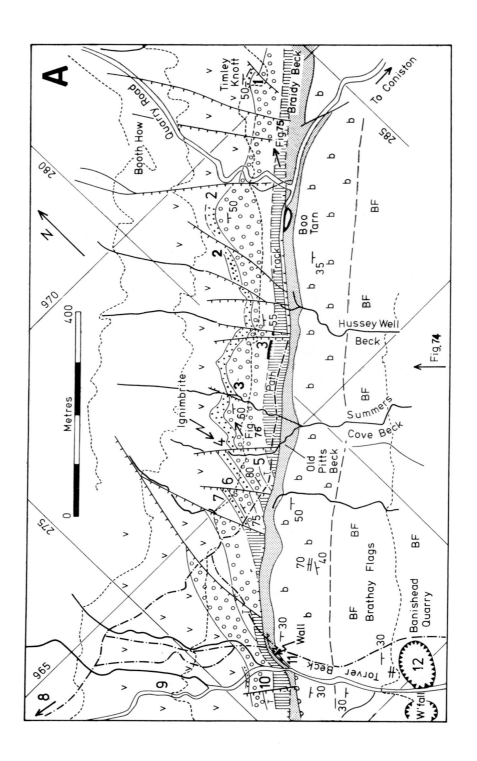

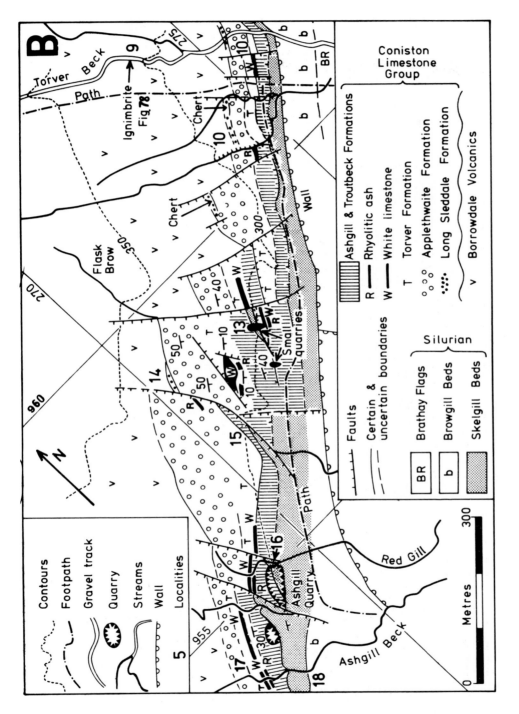

Figure 73 Two maps (A and B) showing details of the geology between Timley Knott and Ashgill Quarry. Understanding of these maps will require a moderately advanced geological standard. The localities are referred to in the text, and the contours are in metres

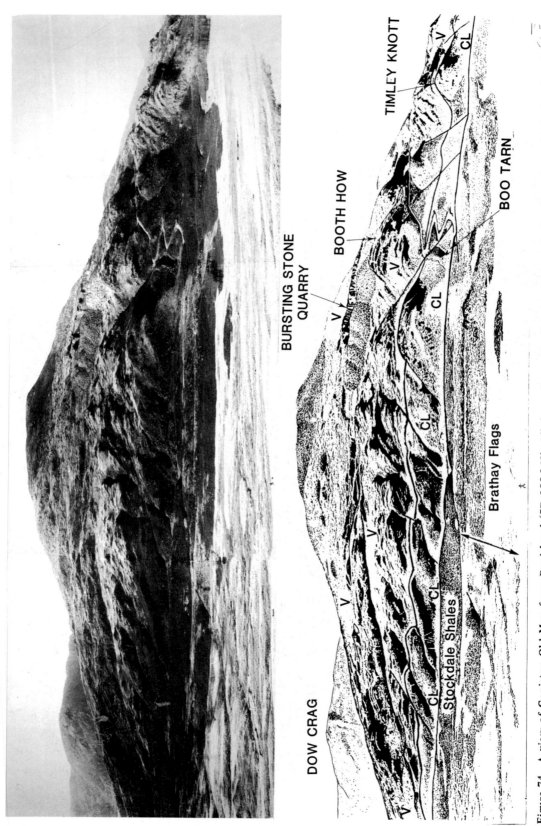

DOW CRAG

BURSTING STONE QUARRY

V

BOOTH HOW

V

V

V

V

V

CL

CL

CL

CL

Stockdale Shales

CL

Brathay Flags

TIMLEY KNOTT

V

CL

BOO TARN

V

N

V

↑

Figure 74 A view of Coniston Old Man from Banishead (SD 288 963). All the rocks are dipping steeply towards the viewpoint and it is therefore difficult to pick out geological structures. However, the oldest volcanic rocks form the summit of Coniston Old Man; bedded tuff is worked in Burstingstone Quarry and the rough crags nearer the viewpoint are in Yewdale Breccia. The Coniston Limestone forms a slab resting on the volcanics at the bottom of the fell, and the Silurian rocks form the lower ground

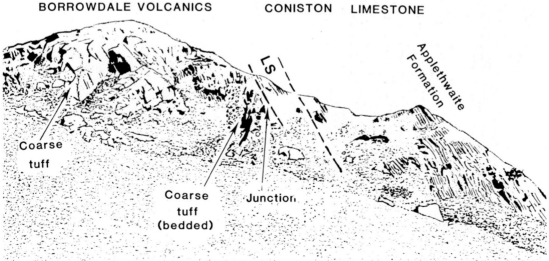

BORROWDALE VOLCANICS   CONISTON   LIMESTONE

LS

Applethwaite Formation

Coarse
tuff

Coarse
tuff
(bedded)

Junction

Figure 75 A view of Timley Knott (locality 1, SD 284 971) from the bottom of the quarry road. The junction between the volcanics and the Coniston Limestone can be readily appreciated from this point (LS, Longsleddale Formation)

Figure 76   The rock-climbers' precipice of Dow Crag (SD 263 978) rising above Goats Water is formed of andesite lava

would suggest. They are as follows with the highest beds first.

A. Applethwaite Formation of the main crag. This shows alternating bands of calcareous mudstone and limestone, the latter tending to form lensoid masses about 20 cm in diameter. There are also interbedded layers of laminated (banded) silt grade sediment, mostly composed of volcanic fragments. There is another small exposure of Applethwaite Beds north-west of the col.

B. Longsleddale Formation. There is a 3 metre gap between the Applethwaite Formation and a tuff-like outcrop which first acquaintance is likely to include with the volcanics. Other localities (2, 3 and 4) however show these beds to be fossiliferous and that they are really a sandstone made up of volcanic fragments. The sandstone is underlain by a well-banded hard silicified rock (chert) which has also yielded fossils and shows that the Longsleddale Formation belongs to the Coniston Limestone Group.

C. Yewdale Breccia. Coarse-grained bedded tuff with angular white fragments of fine-grained rock, and lenticular quartz veins.

2. 282 969. The Coniston Limestone curves round a prominent hill at the top of which the junction between the Applethwaite Formation and underlying tuff-like deposits can be seen. The questions to be considered by those examining this exposure are: (i) Is this deposit a true tuff and part of the Borrowdale Volcanics? (ii) Is it a sandstone made up of grains eroded from the volcanics and in fact the basal bed of the Coniston Limestone Group (the Longsleddale Formation)? Walk 100 metres south-west across the head of a small gully and a tuff-like deposit similar to that of locality 1 will be found. It has thin interbeds and angular fragments of fossiliferous chert, which calls to mind an affinity with the Longsleddale Formation, the lowest division of the Coniston Limestone Group. A few metres to the north-west there are crags of Yewdale Breccia, a coarse tuff with angular fragments and ignimbrite streaks in the matrix.

3. 280 967. A prominent gully exposes White Limestone and Torver mudstone. Interpretation, however, is difficult until the better exposures further south-west have been seen. Further up the gully Applethwaite Beds and fossiliferous 'volcanic sandstones' (Longsleddale Beds) are exposed. Near the bottom of the gully a small path slants uphill. After 100 metres there is an outcrop of chertified Longsleddale Beds with occasional weathered-out (calcareous) concretions. Fifty metres beyond this is the Applethwaite Beds exposure shown on figure 77.

4. 278 966. The outcrops within a 10 metre radius of this locality are of considerable interest. The volcanic rocks adjacent to the Coniston Limestone consist of a thin ignimbrite layer underlain by Yewdale Breccia. The ignimbrite has the characteristic eutaxitic texture of figure 20. It is overlain by what appears to be medium-grained bedded tuff (see locality 2) but the fact that it is fossiliferous with occasional brachiopod shells indicates that it is the lowest member of the Coniston Limestone Group, the Longsleddale Formation (McNamara, 1979). It is a sandstone formed of volcanic rather than quartz fragments and is overlain by the Applethwaite Formation. An obvious fault will be noticed immediately south-west of these exposures, and on the south-west side of the fault the Longsleddale Formation is again seen, this time with a thin layer of conglomerate near the base (a deposit made up of subspherical water-worn pebbles).

5. 279 965. A small rib of cleaved mudstone runs obliquely to the footpath. This is part of the Torver Formation.

6 and 7. 278 965. The sequences here are similar to those at locality 4. Notice that the outcrop of the Skelgill Beds hereabouts truncates bedding and other structures within the Coniston Limestone (figure 73A). This could be interpreted as an unconformity but careful mapping of many outcrops suggests that a more likely interpretation is that of a fault subparallel to the bedding planes in the Skelgill Beds. Such a fault also helps to explain apparent thickness variations within the Skelgill Beds from one area to another.

8. 271 965. Follow the path (Walna Scar road) to the bridge across Torver Beck. Here there are

Figure 77 An exposure of the Applethwaite Formation, Coniston Limestone, at SD 279 966 (see figure 75). Limestone beds (paler) alternate
with calcareous mudstone. The beds are dipping 50° south-east (left to
right) and the cleavage dips 70° north-west

good outcrops of Yewdale bedded tuff with bedding planes tilted to near vertical.

9. 274 963. Follow a path downstream from the bridge until, on the left bank, excellent exposures of columnar jointed ignimbrite will be seen (figure 78). These are cooling joints formed when the rock was still very hot. Close examination will also reveal a eutaxitic texture (see also figure 19).

10. 276 962 to 274 961. Continue downstream until the Coniston Limestone outcrop is reached. This small area extending about 200 metres from Torver Beck to its small tributaries can provide an interesting mapping exercise for undergraduate parties (see localities 16 and 17 for comment on the White Limestone and the Rhyolitic Ash (tuff)).

11. 277 962. The outcrop adjacent to Torver Beck (where the wall crosses) shows the upward passage from Skelgill Beds to Browgill Beds. The Browgill Beds are continuously exposed downstream from here and gradually pass into the striped silty mudstones of the Brathay Flags.

12. 278 960. The Brathay Flags are well seen in Banishead Quarry (figure 23A). They are finely banded rocks with grey mudstone and paler silty layers alternating. They are also strongly cleaved and were quarried for roofing slate.

13. 272 958. Follow the path south-west until some small quarries are seen. The first of these shows an interesting section. In the back of the quarry grey Ashgill Shales are exposed dipping at about 40° south-east. They are followed by black Skelgill Shales rich in graptolites. A north-east–south-west fault through the middle of the quarry cuts off these beds and in the quarry entrance the Torver Formation, White Limestone and Rhyolitic Ash can be seen (see localities 16 and 17).

14. 271 959. The Applethwaite Formation makes a small escarpment and, although exposures are not continuous, the calcareous nature is evident from a number of sink holes (swallow holes or solution holes characteristic of limestone outcrops). It is underlain by the volcanic sandstone of the Longsleddale Formation (see locality 4).

15. 271 957. Walk downstream from locality 14. There is a good exposure of the Rhyolitic Ash at the top of the small waterfall. Further downstream

a narrow slice of steeply dipping Ashgill Shales and graptolitic Skelgill Shales is brought in by a fault.

16. 269 955, Ashgill Quarry. The quarry and its surrounds are of great interest and it is worth spending some time in this area. Outcrops north of the quarry include most of the Coniston Limestone formations described by McNamara. The Applethwaite Formation has the same characteristics as those already described. Above it the Torver fossiliferous mudstones can be seen, followed by the White Limestone and the Rhyolitic Ash. The former is a fairly pure limestone several metres thick, resembling the more calcareous parts of the Applethwaite Formation, and the Rhyolitic Ash (tuff) is a fine-grained, pale, hard rock also several metres thick. It is however difficult to find convincing evidence for the Troutbeck Formation hereabouts (figure 73), but the overlying Ashgill Shales are well exposed in the north-west part of the quarry. The shales are uniform grey with occasional fossils (brachiopods etc.) and split along a strong cleavage inclined steeply to the north-west. The Ordovician–Silurian boundary, a major stratigraphical junction, runs across the quarry floor and rises up the south-east face, so that it can be studied in detail. The lowest Silurian rocks, the Skelgill Shales, are well seen and are black striped mudstones quite different to the Ashgill Shales. They split along bedding planes more easily than along the cleavage and this facilitates observation of the graptolites, which are numerous along certain bedding planes. The Skelgill Shales have been subdivided by R. B. Rickards (see Ingham *et al.*, 1978) and J. E. Hutt (see Hutt, 1974) into ten zones on the basis of the graptolites. At the south-west end of the quarry an obvious fault brings Skelgill Shales against Ashgill Shales.

17. 267 954, Ashgill Beck. This is a classic sequence first described by J. E. Marr and includes most of the Coniston Limestone Group (Marr, 1916b). Ashgill Shales are exposed in the waterfall, with the Troutbeck, White Limestone, Rhyolite Ash and Torver Formations exposed upstream.

18. 268 953. Browgill Beds are to be seen further downstream.

Figure 78   Columnar jointing in the Yewdale ignimbrite, Torver Beck (locality 9, SD 274 963). These joints were formed as hexagonal columns during the cooling of the ignimbrite. Compare with the cracks that form when mud dries

Gb.   Long Crag to Church Beck, Coniston (1 : 25 000 maps SD 29, SD 39)

The outcrops on Long Crag consist of the highest part of the volcanic succession hereabouts, lower formations being exposed further north-west beyond the old copper mines, towards Levers Water. The latter have been described in an excursion guide by G. H. Mitchell (Mitchell, 1970).

To follow the excursion route shown on figure 79 take the footpath that skirts the bottom of the crag (SD 300 980). This approximately follows the Coniston Limestone outcrop which here is badly exposed. Note however that both the Coniston Limestone and graptolitic Skelgill Beds are well exposed in Church Beck at SD 298 978.

1. 300 981. From the bottom of the crag a small path climbs through bracken passing outcrops of breccia and large fallen blocks of eutaxitic ignimbrite, some of which show columnar jointing (see figures 19 and 78).

2. 300 983. The path leads to a steep narrow scree, which is the line of a small fault, with crags exposed on the south side. These crags are of ignimbrite much of which displays columnar joints as shown on figure 78.

3. 299 984. At the top of Long Crag and beyond, Yewdale bedded tuffs are exposed and have a near-vertical dip (figure 80). A descent can be made from here to Church Beck near the Youth Hostel (290 985) from where it is a short walk back to Coniston. There is a great deal of interest in this

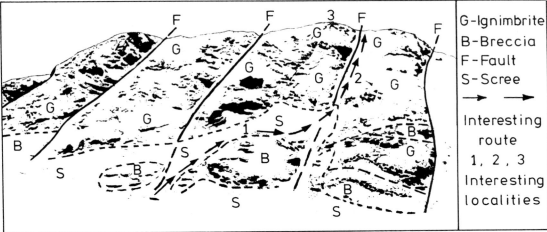

Figure 79   Long Crag, Coniston from Coniston Village, SD 302 979, with a suggested short excursion route shown. From this position it is easy to pick out small faults along which gullies have been eroded. However, the rocks are inclined steeply towards the viewpoint and the boundaries between different rock units are not easily seen. The Coniston Limestone occupies the foreground

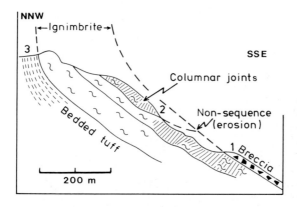

Figure 80   A section across Long Crag, Coniston as represented by Millward (1980) (excursion Gb). Ignimbrite with columnar joints and eutaxitic texture is well seen on this traverse

area also, and an excursion guide has been published by G. H. Mitchell (Mitchell, 1970). For example, at 293 983 where Red Dell Beck crosses the track there is an andesite lava with perlitic cracks. These are small concentric cooling cracks, indicating that the lava was once glassy. A little further down the track (294 982) exposures on an ice-smoothed *roche moutonnée* show vesicular andesite and at 295 981 Yewdale bedded tuff can be seen alongside the track with bedding inclined vertically. Nearer to Coniston, following the track to the north of Church Beck, there is an excellent section in vertical Yewdale Breccia (296 980). This rock consists of angular fragments of variable composition and has a matrix that is often strongly ignimbritic (streaky black fiamme). This suggests that it was hot when deposited.

Gc. Hodge Close, Coniston (NY 316 015)
(1 : 25 000 map NY 30)

There are many quarries between Coniston and Langdale, both working and abandoned, that have exploited the water-lain (volcaniclastic) tuffs characteristic of this part of the volcanic sequence. These are the Lakeland green slates that make such attractive ornamental stone and are to be found on shop fronts and office blocks from Birmingham, England to Birmingham, Alabama. The disused Hodge Close quarry is one of the most accessible of these quarries and can be reached by following a minor road leading off the main Coniston–Ambleside road at 314 998.

A view of the main (south) quarry looking north (figure 81) shows immediately that the strata have been tilted to the vertical and that there are numerous minor contortions in the bedding. There is also a strong cleavage.

Close inspection can only be achieved by following the road to the northern end of the second (north) quarry (NY 317 019) and then descending by a path into the quarry. This leads to a short tunnel and then to the bottom of the water-filled south quarry, a locality to delight the hearts of rock climbers, skin divers, small boys and frogs. Specimens can be examined from here and from the debris in the north quarry which show a great variety of sedimentary structures in water-lain ash deposits, now referred to as volcaniclastic tuffs (figure 82). It will be noticed that the rock splits along well-defined cleavage planes that are at an angle to the bedding, and it is the pattern of the bedding on these cleavage planes that enhances the ornamental value of the rock.

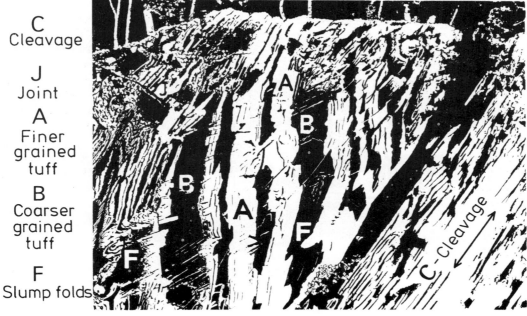

C
Cleavage

J
Joint

A
Finer
grained
tuff

B
Coarser
grained
tuff

F
Slump folds

Figure 81 The northern side of Hodge Close quarry (SY 317 017). Bedded volcaniclastic (water-lain) tuffs have been tilted to vertical and strongly cleaved by the Caledonian earth movements. They display a great variety of sedimentary structures (see figures 16 and 82)

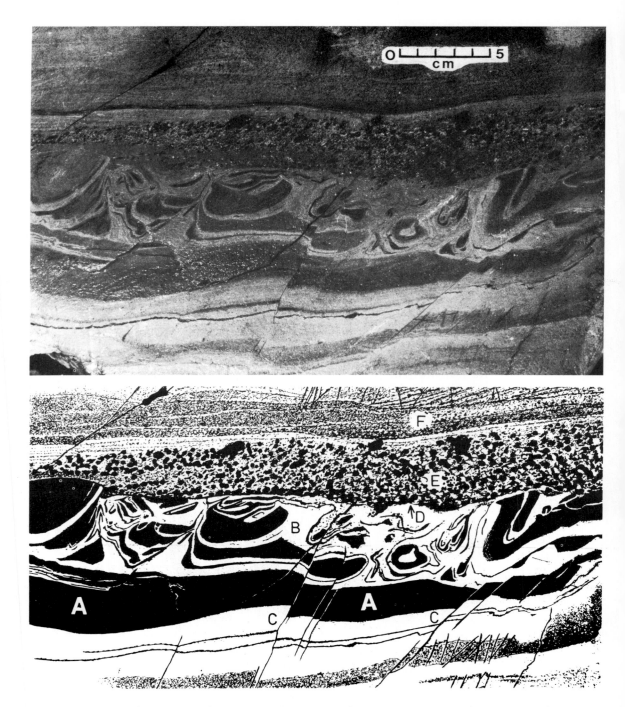

Figure 82 Bedded (volcaniclastic) tuff from Hodge Close (SY 317 017, figure 81). A slumped (convolute bedding) unit is overlain by bedded tuff

References

Bott, M. H. P. (1978). Deep Structure. In *The Geology of the Lake District*, ed. F. Moseley, Occasional Publication No. 3, Yorkshire Geological Society, Leeds, pp. 25-40

Capewell, J. G. (1954). The basic intrusions and an associated vent near Little Mell Fell, Cumberland. *Trans. Leeds geol. Soc.*, **6**, 243-8

Clark, L. (1964). The Borrowdale Volcanic Series between Borrowdale and Wasdale, Cumberland. *Proc. Yorks. geol. Soc.*, **34**, 343-56

Cumberland Geological Society (1982). *The Lake District*. Unwin, London

Firman, R. J. (1978). Intrusions. In *The Geology of the Lake District*, ed. F. Moseley, Occasional Publication No. 3, Yorkshire Geological Society, Leeds, pp. 146-63

Holland, E. G. (1981). *Coniston Copper Mines*. Cicerone Press, Milnthorpe, Cumbria

Hutt, J. E. (1974). The Llandovery graptolites of the English Lake District Pt. 1. *Palaeontogr. Soc. [Monogr.]*, **128**, 1

Ingham, J. K., McNamara, K. J. and Rickards, R. B. (1978). The Upper Ordovician and Silurian rocks. In *The Geology of the Lake District*, ed. F. Moseley, Occasional Publication No. 3, Yorkshire Geological Society, Leeds, pp. 121-45

Jackson, D. (1978). The Skiddaw Group. In *The Geology of the Lake District*, ed. F. Moseley, Occasional Publication No. 3, Yorkshire Geological Society, Leeds, pp. 79-98

Jeans, P. J. F. (1971). The relationship between the Skiddaw Slates and the Borrowdale Volcanics. *Nature: Phys. Sci.*, **234**, 59

Jeans, P. J. F. (1972). The junction between the Skiddaw Slates and Borrowdale Volcanics in Newlands Beck, Cumberland. *Geol. Mag.*, **109**, 25-8

Marr, J. E. (1916a). *The Geology of the Lake District*. Cambridge University Press, Cambridge

Marr, J. E. (1916b). The Ashgillian succession in the tract to the west of Coniston Lake. *Q. Jl. geol. Soc. Lond.*, **71**, 189-204

McNamara, K. J. (1979). The age, stratigraphy and genesis of the Coniston Limestone Group in the southern Lake District. *Geol. J.*, **14**, 41-67

Millward, D. (1979). Ignimbrite volcanism in the Ordovician Borrowdale volcanics of the English Lake District. In *The Caledonides of the British Isles*, Geological Society, London, pp. 629-34

Millward, D. (1980). Three ignimbrites from the Borrowdale Volcanic Group. *Proc. Yorks. geol. Soc.*, **42**, 595-616

Millward, D., Moseley, F. and Soper, N. J. (1978). The Eycott and Borrowdale volcanic rocks. In *The Geology of the Lake District*, ed. F. Moseley, Occasional Publication No. 3, Yorkshire Geological Society, Leeds, pp. 99-120

Mitchell, G. H. (1940). The Borrowdale Volcanic Series of Coniston, Lancashire. *Q. Jl. geol. Soc. Lond.*, **96**, 301-19

Mitchell, G. H. (1956). The geological history of the Lake District. *Proc. Yorks. geol. Soc.*, **30**, 407-63

Mitchell, G. H. (1970). The Lake District. *Geologists' Association Guide No. 2*, Geologists' Association, London, pp. 1-42

Mitchell, G. H., Moseley, F., Firman, R. J., Soper, N. J., Roberts, D. E., Nutt, M. J. C. and Wadge, A. J. (1972). Excursion to the northern Lake District. *Proc. Geol. Ass.*, **83**, 443-70

Moseley, F. (1960). The succession and structure of the Borrowdale Volcanic rocks south-east of Ullswater. *Q. Jl. geol. Soc. Lond.*, **116**, 55-84

Moseley, F. (1964). The succession and structure of the Borrowdale Volcanic rocks north-west of Ullswater. *Geol. J.*, **4**, 127-42

Moseley, F. (1972). A tectonic history of North-West England. *J. geol. Soc. Lond.*, **128**, 561-98

Moseley, F. (1975). Structural relations between the Skiddaw Slates and the Borrowdale Volcanics. *Proc. Cumb. geol. Soc.*, **3**, 127-45

Moseley, F. (1977). Explosion breccias in the Borrowdale volcanics of High Rigg, near Keswick, Cumbria. *Proc. Cumb. geol. Soc.*, **3**, 197-207

Moseley, F. (Ed.) (1978). *The Geology of the Lake District*. Occasional Publication No. 3, Yorkshire Geological Society, Leeds

Moseley, F. (1981). Field meeting to the northern Lake District. *Proc. Yorks. geol. Soc.*, **43**, 395-410

Moseley, F. (1982). Lower Palaeozoic volcanic environments in the British Isles. In *Igneous Rocks of the British Isles*, ed. D. Sutherland, Wiley, Chichester, pp. 39-44

Moseley, F. and Millward, D. (1982). Ordovician volcanicity in the English Lake District. In *Igneous Rocks of the British Isles*, ed. D. Sutherland, Wiley, Chichester, pp. 93-111

Oliver, R. L. (1961). The Borrowdale volcanic and associated rocks of the Scafell area, English Lake District. *Q. Jl. geol. Soc. Lond.*, **117**, 377-417

Roberts, D. E. (1971). Structures of the Skiddaw Slates in the Caldew Valley, Cumberland. *Geol. J.*, **7**, 225-38

Rundle, C. C. (1979). Ordovician Intrusions in the English Lake District. *J. geol. Soc. Lond.*, **136**, 29-38

Shackleton, E. H. (1968). *Lakeland Geology*. Dalesman, Clapham, North Yorkshire

Shackleton, E. H. (1975). *Geological Excursions in Lakeland*. Dalesman, Clapham, North Yorkshire

Shaw, W. T. (1970). *Mining in the Lake Counties*. Dalesman, Clapham, North Yorkshire

Simpson, A. (1967). The stratigraphy and tectonics of the Skiddaw Slates and relationship of the overlying Volcanic Series in part of the Lake District. *Geol. J.*, **5**, 391-418

Soper, N. J. (1970). Three critical localities on the junction of the Borrowdale Volcanic rocks with the Skiddaw Slates in the Lake District. *Proc. Yorks. geol. Soc.*, **37**, 461-93

Soper, N. J. and Moseley, F. (1978). Structure. In *The Geology of the Lake District*, ed. F. Moseley, Occasional Publication No. 3, Yorkshire Geological Society, Leeds, pp. 45-67

Wadge, A. J. (1978). Classification and stratigraphical relationships of the Lower Ordovician rocks. In *The Geology of the Lake District*, ed. F. Moseley, Occasional Publication No. 3, Yorkshire Geological Society, Leeds, pp. 68-78

Wadge, A. J., Harding, R. R. and Darbyshire, D. P. F. (1974). The rubidium strontium age and field relations of the Threlkeld Microgranite. *Proc. Yorks. geol. Soc.*, **40**, 211-22

Ward, J. C. (1876). The geology of the northern part of the English Lake District. *Mem. geol. Surv. U.K.*

Webb, B. C. (1972). N-S trending pre-cleavage folds in the Skiddaw Slates Group of the English Lake District. *Nature: Phys. Sci.*, **235**, 138-40

Index